FIELD HANDLING OF

NATURAL GAS

THIRD EDITION

PETROLEUM EXTENSION SERVICE
The University of Texas at Austin
Austin, Texas

in cooperation with
TEXAS EDUCATION AGENCY
Department of Occupational Education and Technology
Austin

and

COMMITTEE ON VOCATIONAL TRAINING
American Petroleum Institute
Division of Production
Dallas

and

NATURAL GAS PROCESSORS ASSOCIATION
Tulsa

1972

ISBN-0-88698-077-1

TABLE OF CONTENTS

FOREWORD

This is the third edition of the manual *Field Handling of Natural Gas.* The information contained herein updates and is the modern version of the procedures, practices, and equipment described in the earlier editions. The information was collected by a special group headed by H. J. "Hank" Haas, chairman of the Advisory Committee for the School of Gas Technology. The school is operated by Petroleum Extension Service at Kilgore College and is jointly sponsored by the API, Division of Production, and NGPA.

Grateful acknowledgment is made to the following people for their work in compiling and reviewing the material in this manual.

Henry J. Haas, HNG Petrochemicals, Inc.

Ronald E. Cannon, Executive Director,.Natural Gas Processors Association

Walter Huffman, Black Sivalls and Bryson Company

Paul Morgan, C-E Natco

Hubert Pringle, Amoco Production Company

Millard Clegg, Humble Oil and Refining Company

Willard Sutton, Houston Natural Gas Company

Jim Caldwell, Cooper-Bessemer Company

Jason Troth, Solar Division, International Harvester Company

Curtis F. Kruse, Petroleum Extension Service

Robert N. Hastings, who has retired from Amoco Production Company, worked to edit the written material submitted by those listed above. The cooperation by all concerned in the preparation of this manual that is to be used as a study guide for the School of Gas Technology is acknowledged.

W. E. Boyd, Director
Petroleum Extension Service
The University of Texas at Austin

Austin, Texas
July, 1972

PREFACE

It is the purpose of this manual to present fundamental concepts and workable practices for the handling of natural gas. The manual has been prepared primarily to aid the field employee who deals regularly with the problems of producing, transporting, and processing gas.

The special review committee appointed to validate this third edition has been mindful of this purpose and has endeavored to include positive, useful information, while minimizing highly technical material that would not contribute to the productive capability of the field employee.

The manual does not cover all materials, practices, procedures, or equipment used in the field handling of natural gas. Neither should this book be interpreted as indicating a preference by the committee for one specific manner of operation. The reader must be guided by his own experience in dealing with variances in local conditions and requirements.

It has been necessary and desirable to include photographs and drawings describing equipment that may be easily identified as to manufacturer. The committee did not intend by these inclusions to infer in any way a preference or practice that is limited to equipment manufactured by any specific company.

Henry J. Haas, Chairman
API-NGPA Validation Committee

I

INTRODUCTION

Natural gas has been used commercially as a fuel for 150 years in America and for centuries in China. The production, processing, and distribution of natural gas has become an important segment of our domestic economy and is a major factor in the world's energy markets.

Since its discovery in the United States, reported to be at Fredonia, New York, in 1821, natural gas has been used as a fuel in areas immediately surrounding the gas fields. In the 1920s and 1930s, a few long pipelines from 22 to 24 in. in diameter, operating at 400 to 600 psi, were installed to transport gas to industrial areas remote from the field sources.

However, as late as the 1930s produced natural gas was flared and blown to the air in large volumes. When gas accompanied crude oil, the gas had to find a market or be flared and, in the absence of effective conservation practices in earlier years, oil-well gas was often flared in huge quantities. Consequently, gas production at that time was often short-lived, and gas could be purchased for as little as one or two cents per 1,000 cu ft in the field.

A combination of factors, including the low prices for a superior fuel, assurance of continuing supply of gas by strict conservation practices and new discoveries, and satisfactory financial returns to pipeline companies contributed to the fantastic growth of the natural gas industry in recent years.

The modern natural gas industry began immediately following World War II when a number of long-distance pipelines were constructed to serve markets in the populated areas of the country. By that time, advances in welding and manufacture of pipe permitted pressures up to 1,000 psi and diameters up to 30 in. Today virtually every area of the United States is served by natural gas. Figure 1.1 is a map of gas transmission lines at the end of 1969, as issued by the Federal Power Commission. Figure 1.2

is a graph showing growth of natural gas production since 1946, together with projections of requirements to 1980.

Figure 1.3 is a graphic representation of major energy sources for selected years since 1945, showing the relative growth of natural gas consumption and its inroads into markets, largely at the expense of coal. Nuclear power, once touted as a major energy source for electric power generation, has yet to show significant use due to unexpected problems and costs of construction. Thus, natural gas is firmly established as the preferred fuel in nearly every application, with natural gas and gas liquids supplying about 50 percent of the total U.S. consumption of petroleum hydrocarbon energy and 35 percent of the total energy requirements of the nation.

At the present time, the nation is facing an acute crisis in gas supply, which has been brought about by a combination of two factors: (1) the low field prices for natural gas under FPC regulations and (2) extreme demands for clean-burning, low-sulfur fuels, as required by air pollution regulations in the major population centers. As a result, the gas industry's principal problem is that of meeting the prodigious demands that will be imposed during the seventies. To illustrate the magnitude of the problem, it is estimated that consumption of natural gas and gas liquids during the 1970s will equal all of the gas and gas liquids consumed prior to 1969 (fig. 1.4).

Principal reserves of natural gas are found along the Gulf Coast, through the Mid-Continent, and on the eastern slope of the Rocky Mountains. Large new reserves are waiting to be produced from Alaska's North Slope. In addition, the United States will probably require large volumes of imported liquefied natural gas to meet projected demands. On the basis of 1970 consumption and reserves, we have enough natural gas for about 12 years and enough oil

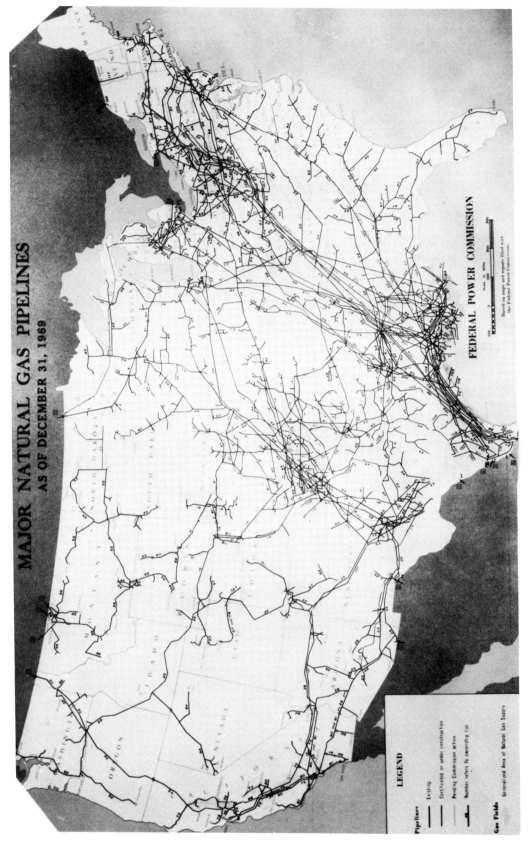

Figure 1.1. Major Natural Gas Pipelines in the United States

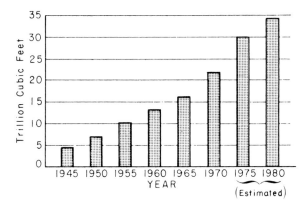

Figure 1.2. Annual Gas Utilization in the United States (US Bureau of Mines Statistics)

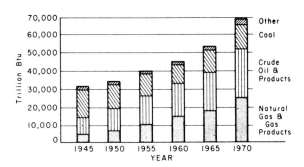

Figure 1.3. Energy Consumption in the United States by Source (US Bureau of Mines Statistics)

for about 10 years, assuming that no new deposits of either were to be discovered. In that year the 290 trillion cubic feet of gas estimated recoverable reserves for the United States will exceed the fuel value of the 40 billion bbl of crude oil reserves. New deposits will be discovered, of course—perhaps greater ones than ever before—but the fact remains that natural gas is not inexhaustible in the earth, and it is irreplaceable. Exploration, discovery, production, processing, and handling of these tremendous energy reserves will require the best of the industry's people in the years to come.

The facilities normally operated in the handling of gas in the field are those required to condition the gas to make it marketable—the removal of impurities, water, and excess hydrocarbon liquids and the control of delivery pressure. The last operation will involve the use of pressure-reducing regulators or, more often, compressors to raise the pressure.

Natural gas processing plants are usually designed to remove certain valuable products over and above

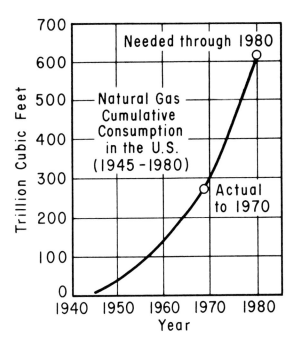

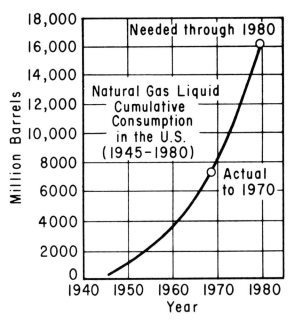

Figure 1.4. Natural Gas and Liquids Consumption. Data shown are actual through 1970, estimated for 1980, and total production prior to 1970.

those needed to make the gas marketable, that is, natural gasoline, butane, propane, ethane, and even methane in some instances. Plants may also be designed to recover elemental sulfur from the hydrogen sulfide gas removed from wellhead gas. Another

3

function of plants is to separate the liquid hydrocarbons recovered into various mixtures or pure products by the use of fractionating columns.

Plants nearly always incorporate in their processes many of the functions ordinarily performed by field facilities. Dehydration and H_2S removal are nearly always part of the plant operation. The equipment used for these processes is essentially that described in chapters on "Hydrates," "Dehydration of Natural Gas," and "Miscellaneous Gas Conditioning."

II

CHARACTERISTICS OF NATURAL GAS

Natural gas is a mixture of hydrocarbon gases along with some impurities that are the result of decomposed organic material. The impurities found also include water vapor and heavier hydrocarbons. When raw natural gas is withdrawn from the underground reservoirs to supply energy demands, these impurities are considered objectionable and are usually removed by various processing schemes. The hydrocarbon gases normally found in natural gas are methane, ethane, propane, butanes, pentanes, and small amounts of hexanes, heptanes, octanes, and the heavier gases. Usually the propane and heavier fractions are removed for additional processing because of their high market value as gasoline-blending stock and chemical plant raw feedstock. What usually reaches the transmission line for sale as natural gas is mostly a mixture of methane and ethane with some small percentage of propane. Methane is usually the largest percentage (95 to 98 percent).

COMPOSITION OF NATURAL GAS

Natural gas has been defined as a mixture of hydrocarbon gases and impurities. There is no one composition or mixture that can be referred to as natural gas. Each gas stream produced has its own composition. Even two gas wells from the same reservoir may have different compositions. Examples of some typical natural gas streams are provided in table 2.1 to show the range of composition that is naturally produced.

Well stream no. 1 is typical of an associated gas; that is, gas produced with crude oil. Well streams no. 2 and no. 3 are typical low-pressure and high-pressure gases of the nonassociated type. Not only is there a wide variety of natural gas compositions, but each gas

Table 2.1

TYPICAL NATURAL GAS ANALYSES

Component	Well No. 1 Mol Percent	Well No. 2 Mol Percent	Well No. 3 Mol Percent
Methane	27.52	71.01	91.25
Ethane	16.34	13.09	3.61
Propane	29.18	7.91	1.37
i-Butane	5.37	1.68	0.31
n-Butane	17.18	2.09	0.44
i-Pentane	2.18	1.17	0.16
n-Pentane	1.72	1.22	0.17
Hexane	0.47	1.02	0.27
Heptanes and Heavier	0.04	0.81	2.42
Carbon Dioxide	0.00	0.00	0.00
Hydrogen Sulfide	0.00	0.00	0.00
Nitrogen	0.00	0.00	0.00
Total	100.00	100.00	100.00

Note: Production from many wells will contain small quantities of carbon dioxide, hydrogen sulfide, and nitrogen.

stream produced from a natural gas reservoir can change composition as the reservoir is depleted. The well analyses should be checked periodically since it may be necessary to change the equipment used for production in order to satisfy the new composition of the gas.

Natural gas is normally thought of as being a mixture of straight-chain or paraffin hydrocarbon gases. We do, however, occasionally find cyclic and aromatic hydrocarbon gases (cyclic compounds) in the mixture. The terms *straight chain* and *cyclic* refer to the molecular structure. The compounds previously referred to as methane, ethane, propane, butanes, pentanes, and so forth are straight-chain or paraffinic compounds. Molecular structure of these and the cyclic compounds are shown for comparison in figure 2.1.

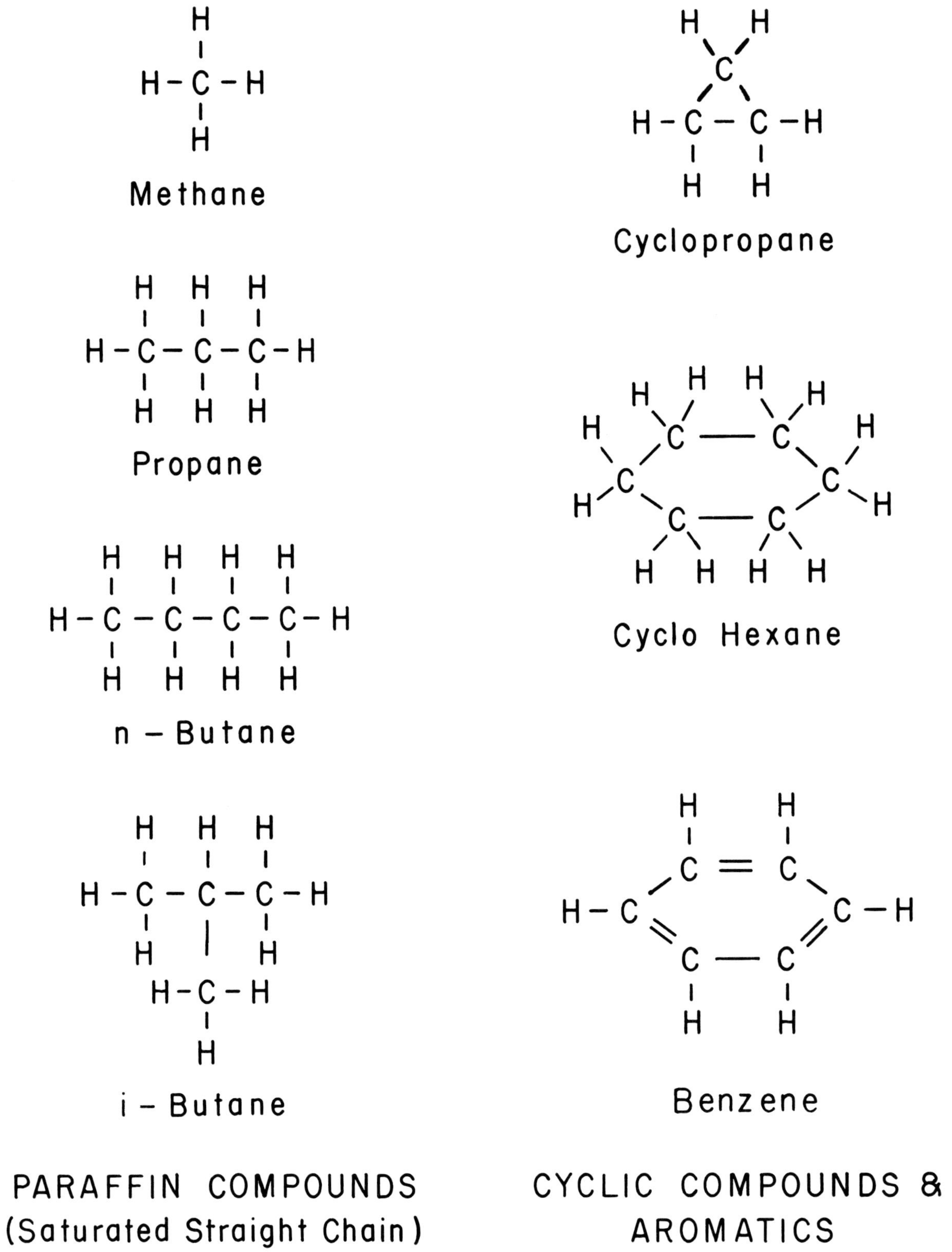

H
|
H–C–H
|
H

Methane

H   H   H
| | |
H–C–C–C–H
| | |
H   H   H

Propane

H   H   H   H
| | | |
H–C–C–C–C–H
| | | |
H   H   H   H

n – Butane

H   H   H
| | |
H–C–C–C–H
| | |
H | H
H–C–H
|
H

i – Butane

PARAFFIN COMPOUNDS
(Saturated Straight Chain)

H   H
\   /
C
/ \
H–C   C–H
| |
H   H

Cyclopropane

H   H
| |
H   C — C   H
\ /   \ /
C   C
H /   \ H
C   C
/ \   / \
H   C — C   H
| | | |
H   H   H   H

Cyclo Hexane

H   H
| |
C = C
/   \
H–C   C–H
\   /
C = C
| |
H   H

Benzene

CYCLIC COMPOUNDS &
AROMATICS

Figure 2.1. Hydrocarbon Gas Molecule Structures

6

Impurities are usually found in natural gas and must be removed because they cause difficulties in handling and processing. Such components as hydrogen sulfide, carbon dioxide, mercaptans, water vapor, noncombustible gases (such as nitrogen and helium), pentanes, and the heavier hydrocarbons are generally considered as impurities since the processed natural gas is usually burned as fuel and these compounds may cause extremely unreliable and hazardous combustion conditions for the consumer. Removal of hydrogen sulfide, which is very poisonous, carbon dioxide, and water vapor eliminates the problems of toxicity, corrosion, and hydrate formation in transmission and distribution systems. Market value of some of these impurities may make it attractive for the producer to remove and market them separately, thereby making the processing of the natural gas more economical. When the pentanes and heavier hydrocarbons are recovered, they provide a good source of blending stocks for the production of gasoline. With the growth of the worldwide market for liquefied petroleum gas (LPG), additional processes for higher recovery of the propane and butane fractions, which make up LPG, have been developed. These fractions are valuable for the petrochemical industry raw stocks.

PHYSICAL PROPERTIES OF NATURAL GAS

Since natural gas is a mixture of hydrocarbon compounds and because this mixture is varied in types of compounds as well as the relative amounts, so will the overall or combined physical properties vary. The overall physical properties of a natural gas are indicators of the behavior of the gas under various processing conditions, and it is therefore important to be able to establish these physical properties. In order to do this, the analysis or composition of the gas must be determined first. Once the composition is known, the various physical properties can be determined by using the physical properties of each pure component in the mixture. Physical properties that are most useful in natural gas processing are molecular weight, freezing point, boiling point, density, critical temperature, critical pressure, heat of vaporization, and specific heat. Table 2.2 is a tabulation of physical constants of paraffin hydrocarbons and other components of natural gas taken from the NGPSA *Engineering Data Book*.

TERMINOLOGY

Gauge Pressure

Gauge pressure (psig) is that pressure indicated by a pressure gauge and is the pressure above or below that of the atmospheric pressure at the point of measurement. Most pressure-measuring devices use a Bourdon tube, which is a curved hollow chamber that tends to straighten out as the internal pressure increases. Atmospheric pressure is applied to the outside of the Bourdon tube; so actually the movement of the tube is the result of the differential between the internal pressure and the external pressure. The average atmospheric pressure at sea level is 14.70 psi or 29.92 in. of mercury. Gauge pressure is reported as pounds per square inch gauge (psig).

Absolute Pressure

The absolute pressure (psia) must be used in all calculations involving volume and pressure relationships. To obtain the absolute pressure, the pressure of the atmosphere must be added to the gauge pressure. To be entirely accurate, the atmospheric pressure on the gauge at the time the gauge pressure is read should be determined, but for most calculations the average sea-level pressure of 14.7 psia can be added. In higher altitudes, a lower atmospheric pressure should be added. Absolute pressure is reported as pounds per square inch absolute (psia).

Vapor Pressure

When a liquid evaporates into a space of limited dimensions, the space will become filled with the vapor that is formed. As vaporization proceeds, the number of molecules in the vapor state will increase and cause an increase in the pressure exerted by the vapor. The pressure exerted by a gas, or vapor, is due to the impacts of its component molecules against the confining surfaces. Since the original liquid surface forms one of the walls confining the vapor, there will be a continual series of impacts against the liquid surface by the molecules in the vapor state. The number of such impacts will be dependent on the pressure exerted by the vapor. However, when one of the gaseous molecules strikes the liquid surface, it comes under the influence of the attractive forces of the densely aggregated liquid molecules, is held there, and forms a part of the liquid once more. This phenomenon, the reverse of vaporization, is known as condensation. The rate of condensation is determined by the number of molecules striking the liquid

Table 2.2

PHYSICAL CONSTANTS OF PARAFFIN HYDROCARBONS AND OTHER COMPONENTS OF NATURAL GAS

PHYSICAL CONSTANTS OF PARAFFIN HYDROCARBONS AND OTHER COMPONENTS OF NATURAL GAS

NGPA Publication 2145 - 71[1]

Component	Notes	Methane	Ethane	Propane	Iso-Butane	N-Butane	Iso-Pentane	N-Pentane	N-Hexane	N-Heptane	N-Octane	N-Nonane	N-Decane	Carbon Dioxide	Hydrogen Sulfide	Nitrogen	Oxygen	Air	Water
Molecular Weight	*	16.043	30.070	44.097	58.124	58.124	72.151	72.151	86.178	100.205	114.232	128.259	142.286	44.010	34.076	28.013	31.999	28.964	18.015
Boiling Point @ 14.696 psia, °F		-258.69	-127.48	-43.67	10.90	31.10	82.12	96.92	155.72	209.17	258.22	303.47	345.48	-109.3[2]	-76.6[24]	-320.4[2]	-297.4[2]	-317.6[2]	212.0
Freezing Point @ 14.696 psia, °F		-296.46[d]	-297.89[d]	-305.84[d]	-255.29	-217.05	-255.83	-201.51	-139.58	-131.05	-70.18	-64.28	-21.36	—	-117.2[7]	-346.0[24]	-361.8[24]	—	32.0
Vapor Pressure @ 100°F, psia		(5000)	(800)	190	72.2	51.6	20.44	15.570	4.956	1.620	0.537	0.179	0.0597	—	394.0[6]	—	—	—	0.9492[12]
Density of Liquid @ 60°F & 14.696 psia																			
Specific Gravity @ 60°F/60°F	a,b	0.3[i]	0.3564[h]	0.5077[h]	0.5631[h]	0.5844[h]	0.6247	0.6310	0.6640	0.6882	0.7068	0.7217	0.7342	0.827[h,6]	0.79[h,6]	0.808[m,3]	1.14[m,3]	0.856[m,3]	1.000
°API	*, a,b	340[i]	265.5[h]	147.2[h]	119.8[h]	110.6[h]	95.0	92.7	81.6	74.1	68.7	64.6	61.2	39.6[h]	47.6[h]	43.6[m]	-7.4[m]	33.8[m]	10.0
Lb/gal @ 60°F, wt in vacuum		2.5[i]	2.971[h]	4.233[h]	4.695[h]	4.872[h]	5.208	5.261	5.536	5.738	5.893	6.017	6.121	6.89[h]	6.59[h]	6.74[h]	9.50[m]	7.14[m]	8.337
Lb/gal @ 60°F, wt in air	c	2.5[i]	2.962[h]	4.223[h]	4.686[h]	4.865[h]	5.199	5.251	5.526	5.728	5.883	6.008	6.112	6.89[h]	6.58[h]	6.73[m]	9.50[m]	7.13[m]	8.328
Density of Gas @ 60°F & 14.696 psia																			
Specific Gravity, Air = 1.00, ideal gas	*	0.5539	1.0382	1.5225	2.0068	2.0068	2.4911	2.4911	2.9753	3.4596	3.9439	4.4282	4.9125	1.5195	1.1765	0.9672	1.1048	1.0000	0.6220
Lb/M cu ft, ideal gas	*	42.28	79.24	116.20	153.16	153.16	190.13	190.13	227.09	264.05	301.01	337.98	374.94	115.97	89.79	73.82	84.32	76.32	47.47
Volume Ratio @ 60°F and 14.696 psia																			
Gal/lb mol	*	6.4[i]	10.12[h]	10.42[h]	12.38[h]	11.93[h]	13.85	13.71	15.57	17.46	19.39	21.32	23.24	6.38[h]	5.17[h]	4.16[m]	3.37[m]	4.06[m]	2.16
Cu ft gas/gal liquid, ideal gas	*	59[i]	37.5[h]	36.43[h]	30.65[h]	31.81[h]	27.39	27.67	24.38	21.73	19.58	17.80	16.33	59.5[h]	73.3[h]	91.3[m]	112.7[m]	93.5[m]	175.6
Gas vol/liquid vol, ideal gas	*	443[i]	280.5[h]	272.51[h]	229.30[h]	237.98[h]	204.93	207.00	182.37	162.56	146.45	133.18	122.13	444.8[h]	548.7[h]	682.7[m]	843.2[m]	699.5[m]	1313.8
Critical Conditions																			
Temperature, °F	*	-116.63	90.09	206.01	274.98	305.65	369.10	385.7	453.7	512.8	564.22	610.68	652.1	87.9[23]	212.7[17]	-232.4[24]	-181.1[17]	-221.3[7]	705.6[17]
Pressure, psia	*	667.8	707.8	616.3	529.1	550.7	490.4	488.6	436.9	396.8	360.6	332	304	1071[17]	1306[17]	493.0[24]	736.9[24]	547[7]	3208[17]
Gross Heat of Combustion @ 60°F	p																		
Btu/lb liquid	*	—	22,214[4]	21,513[d]	21,091[d]	21,139[d]	20,889	20,928	20,784	20,681	20,604	20,544	20,494	—	—	—	—	—	—
Btu/lb gas	*	23,885	22,323	21,665	21,237	21,298	21,040	21,089	20,944	20,840	20,762	20,701	20,649	—	—	—	—	—	—
Btu/cu ft, ideal gas	*	1009.7	1768.8	2517.5	3252.7	3262.1	4000.3	4009.6	4756.2	5502.8	6249.7	6996.5	7742.1	—	637[16]	—	—	—	—
Btu/gal liquid	*	—	65,998[d]	91,065[d]	99,022[d]	102,989[d]	108,790	110,102	115,060	118,668	121,419	123,613	125,444	—	—	—	—	—	—
Cu ft air to burn 1 cu ft gas - ideal gas	*	9.54	16.70	23.86	31.02	31.02	38.18	38.18	45.34	52.50	59.65	66.81	73.97	—	7.16	—	—	—	—
Flammability Limits @ 100°F & 14.696 psia	p																		
Lower, vol % in air	*	5.0	2.9	2.1	1.8	1.8	1.4	1.4	1.2	1.0	0.96	0.87[s]	0.78[s]	—	4.30[2]	—	—	—	—
Upper, vol % in air	*	15.0	13.0	9.5	8.4	8.4	(8.3)	8.3	7.7	7.0	7.0	2.9	2.6	—	45.50	—	—	—	—
Heat of Vaporization @ 14.696 psia																			
Btu/lb @ boiling point	*	219.22	210.41	183.05	157.53	165.65	147.13	153.59	143.95	136.01	129.53	123.76	118.68	238.2[14]	235.6[7]	87.8[14]	91.6[14]	92[3]	970.3[1]
Specific Heat @ 60°F & 14.696 psia																			
Cp gas — Btu/lb, °F ideal gas	*	0.5266	0.4097	0.3881	0.3872	0.3867	0.3827	0.3883	0.3864	0.3875	(0.3876)	0.3840	0.3835	0.1991[13]	0.238[14]	0.2482[13]	0.2188[13]	0.2400[13]	0.4446[13]
Cv gas — Btu/lb, °F ideal gas	*	0.4027	0.3436	0.3430	0.3530	0.3525	0.3552	0.3608	0.3633	0.3677	0.3702	0.3685	0.3695	0.1539	0.1797	0.1773	0.1567	0.1714	0.3343
N = Cp/Cv	*	1.308	1.192	1.131	1.097	1.097	1.078	1.076	1.063	1.054	1.047	1.042	1.038	1.293	1.325	1.400	1.396	1.400	1.330
Cp liquid — Btu/lb, °F	*	—	0.9256	0.5920	0.5695	0.5636	0.5353	0.5441	0.5332	0.5283	0.5239	0.5228	0.5208	—	—	—	—	—	1.0009[7]
Octane Number																			
Motor clear		—	+.05[f]	+1.6[f]	97.6	89.6[f]	90.3	62.6[f]	26.0	0.0	—	—	—	—	—	—	—	—	—
Research clear		—	+1.6[f]	+1.8[f]	+0.10[f]	93.8[f]	92.3	61.7[f]	24.8	0.0	—	—	—	—	—	—	—	—	—
Refractive Index n_D @ 68°F		—	—	—	—	1.3326[h]	1.35373	1.35748	1.37486	1.38764	1.39743	1.40542	1.41189	—	—	—	—	—	1.3330[8]

NOTES

a. Air saturated hydrocarbons.
b. Absolute values from weights in vacuum.
c. The apparent values for weight in air are shown for users' convenience. All other mass data in this table are on an absolute mass (weight in vacuum) basis.
d. At saturation pressure (triple point).
f. The + sign and number following signify the octane number corresponding to that of 2,2,4 trimethylpentane with the indicated number of ml of TEL added.
h. Saturation pressure and 60°F.
i. Apparent value for methane at 60°F.
j. Average value from octane numbers of more than one sample.
m. Density of liquid, gm/ml at normal boiling point.
n. Heat of sublimation.
p. Gross heat on dry basis at 60°F and 14.696 psia. To convert to water saturation basis, multiply by 0.9825.
s. Extrapolated to room temperature from higher temperature.
* Calculated values. 1969 atomic weights used. See "Constants for Use in Calculations."
( ) Estimated values.

REFERENCES

1. Values for hydrocarbons were selected or calculated from API Project 44 and are identical to or consistent with ASTM DS 4A "Physical Constants of Hydrocarbons C1 - C10," 1971, American Society for Testing Materials, 1916 Race Street, Philadelphia.
2. International critical tables.
3. Hodgman, Handbook of Chemistry & Physics, 31 edition (1949).
4. West, J. R., Chemical Engineering Progress, 44, 287 (1948).
5. Jones, Chemical Review, 22, 1 (1938).
6. Sage & Lacey, API Research Project 37, Monograph (1955).
7. Perry, Chemical Engineers Handbook, 4th edition (1963).
8. Mattieson and Hanna, Oil and Gas Journal, 41, No. 2, 33 (1942).
9. Keenan & Keyes, Thermodynamic Properties of Air (1947).
12. Keenan & Keyes, Thermodynamic Properties of Steam (Twenty-ninth Printing 1956).
13. American Petroleum Institute, Project 44.
14. Dreisbach, Physical Properties of Chemical Compounds, American Chemical Society, 1961.
16. Maxwell, J. B., Data Book on Hydrocarbons, Van Nostrand Co. (1950).
17. Kobe, K. A. & R. E. Lynn, Jr., Chemical Review, 52, 117-236 (1953).
23. Din, "Thermodynamic Functions of Gases," Butterworths (1956).
24. Thermodynamic Research Center Data Project, Texas A&M University, (formerly MCA Research Project).

surface per unit time, which in turn is determined by the pressure of the vapor. When a liquid evaporates into a limited space, two opposing processes are in operation. The process of vaporization tends to change the liquid into the gaseous state; the process of condensation tends to change the gas formed by vaporization back into the liquid state. The rate of condensation is increased as vaporization proceeds, thus increasing the pressure of the vapor. Ultimately, if sufficient liquid is present, the processes will continue until the rate of condensation equals the rate of vaporization. When this condition is reached, a dynamic equilibrium is established. Since the formation of new vapor is compensated by condensation, the pressure of the vapor will remain unchanged. If the pressure of the vapor is changed in either direction from its equilibrium value, it will adjust itself and return to the equilibrium condition. The pressure exerted by the vapor at such equilibrium conditions is termed the vapor pressure of the liquid. All materials exhibit a definite vapor pressure of greater or lesser degree at any temperature above absolute zero. In general, as molecular weight increases, the vapor pressure decreases. The magnitude of the vapor pressure is not dependent upon the amount of liquid present, as long as some liquid is present. Neither is the magnitude dependent upon the amount of liquid surface area present. Vapor pressure is entirely dependent on the magnitudes of the maximum potential energies of attraction, which must be overcome in vaporization.

Partial Pressure and Pure-Component Volume

In a mixture of different gases, the molecules of each component gas are distributed throughout the entire volume of the containing vessel, and the molecules of each component gas contribute by their impacts to the total pressure exerted by the entire mixture. The total pressure is equal to the sum of the pressures exerted by the molecules of each component gas. These statements apply to all gases, whether or not their behavior is ideal. In a mixture of ideal gases, the molecules of each component gas behave independently as though they alone were present in the container. Before considering the actual behavior of the gaseous mixtures, it is necessary to define two terms commonly used, namely, *partial pressure* and *pure-component volume.* By definition, the partial pressure of a component gas, which is present in a mixture of gases, is the pressure that would be exerted by that component gas if it alone were present in the same volume and at the same temperature as the mixture. By definition, the pure-component volume of a component gas, which is present in a mixture of gases, is the volume that would be occupied by that component gas if it alone were present at the same pressure and temperature as the mixture.

The kinetic theory of the constitution of gases is based on the phenomenon that many properties of gaseous mixtures are additive. The additive nature of partial pressures is expressed by Dalton's law, which states that the total pressure exerted by a gaseous mixture is equal to the sum of the partial pressures, that is:

$$P = p_a + p_b + p_c + p_d + \ldots$$

Where P is the total pressure; P_a, P_b, P_c, etc., are the partial pressures of the component gases.

Similarly, the additive nature of pure-component volumes is given by the law of Amagat or Leduc's law, which states that the total volume occupied by a gaseous mixture is equal to the sum of the pure-component volumes, that is:

$$V = v_a + v_b + v_c + \ldots$$

Where V is the total volume of the mixture, and v_a, v_b, and v_c, and so forth are the pure-component volumes of the component gases.

Pressure Base

An arbitrarily specified standard pressure is established by various states and countries as a *pressure base.* This furnishes a convenient standard that can be used to compare different quantities of gas when expressed in terms of volume. Natural gas is produced and sold by volume in the United States and several other countries; therefore, before the volumes of various gases can be compared, the pressure base of each volume must be known. All volume measurements are mathematically converted to the established pressure base before comparison. Some states in the United States have adopted 14.4 psi as their pressure base. Others have adopted pressure bases of 14.7 psi and 15.2 psi. Coupled with the pressure base should be a standard temperature since volume is affected by temperature as well as pressure. All measurement in the United States is referred to a standard temperature of 60 F. Consequently, specification of the pressure base is important when making volume measurement comparisons.

Gas Density and Specific Gravity

The density of a gas is usually expressed as the weight in pounds per cubic foot. Unless specified otherwise, the volumes are at the standard condition of 60 F and 14.7 psia. On this basis, air has a normal density of 0.0763 lb/cu ft. Specific gravity is the ratio of a gas density to the density of air at the same conditions of temperature and pressure. Specific gravity is an important factor in gas measurement.

Liquid Density and Specific Gravity

The density of a liquid is usually expressed as the weight in pounds per cubic foot, similar to the density of gas. However, the specific gravity of liquids is different. In the case of liquids (including hydrocarbons, liquified natural gas, liquid propane, etc.), specific gravity is the ratio of a liquid's density to that of water at 4 C (39.2 F). The density of water at 4 C is 62.43 lb/cu ft.

There is another gravity term that is used with hydrocarbon liquids, and this is API gravity. A special gravity scale was adopted by the American Petroleum Institute for expression of the petroleum products. This scale is defined as follows:

$$\text{Degrees API} = \frac{141.5}{G} - 131.5$$

G is the liquid's specific gravity at 60 F referenced to that of water at 60 F; thus a liquid that has the same density as water at 60 F, i.e., specific gravity of 1.0, will have an API gravity of 10° API. The gravity of a liquid in degrees API is determined by its density at 60 F and is independent of temperature. Readings of API graduated hydrometers at temperatures other than 60 F must be corrected for temperature so as to give the value at 60 F.

Temperatures

The Fahrenheit temperature scale is almost universally used in the gas industry; however, it may be necessary to become more conversant with the centigrade system as it is more universally used throughout the world for temperature measurements in all other industries. The centigrade temperature scale has its zero established at the freezing point of water and its 100 mark at the atmospheric (sea-level) boiling point of water. These were the easiest and therefore most common reference points for the scientists and were used as calibration points for the early thermometers. The Fahrenheit scale was devised using the same reference points except that the freezing point of water is 32 and the boiling point of water occurs at 212.

Absolute temperature came into use when scientists began expanding their investigations of matter and found that correlations could be established using temperature as a variable provided the temperature was expressed in absolute terms; that is, with reference to an absolute zero. By definition, absolute zero is the temperature at which there is very little or no movement taking place within the atoms of matter. In the centigrade scale, this is 273.1 degrees below 0 C and is referred to as the Kelvin scale. In other words, 0 C corresponds to 273.1 K. For the Fahrenheit scale, absolute zero is 460 degrees below 0 F. This absolute scale is referred to as the Rankine scale. Thus:

$$X \text{ C} = (X + 273.1) \text{ K (Kelvin)}$$

and

$$X \text{ F} = (X + 460) \text{ R (Rankine)}$$

Figure 2.2 gives a comparison of the absolute centigrade, and Fahrenheit scales. In the use of gas laws, care must be taken to use the absolute scale for temperature and to use consistent units for the expressions of both the variable and constant terms.

The measurement of temperature may be obtained with liquid-filled glass tubes referred to as thermometers. The early temperature-measuring instruments were limited in their ranges by the materials available and have since been expanded as new methods and materials were developed. Today a wide choice of instruments to measure temperature is available, ranging from liquid-filled thermometers to thermocouples to optical pyrometers. Typical examples of these instruments and their normal range of usage are given in chapter nine.

Mass and Matter

All matter has mass and weight even though it may be very small. The gases that make up natural gas have different masses and weights. Since the word weight has become entrenched in engineering literature as synonomous with mass, it has become common practice to use the weight of a material rather than the mass as a measure of quantity. Mass and weight are equal only at sea level, but the variation of weight on the earth's surface is negligible in ordinary engineering work.

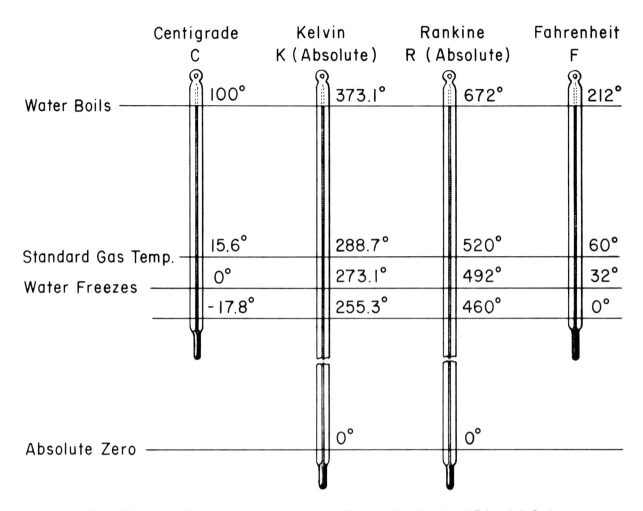

Figure 2.2. Comparison of Absolute Temperature Scales to Centigrade and Fahrenheit Scales

Molecular Weight

Natural gas is a mixture of several different gases that have different characteristics, one of which is structure, as shown before in figure 2.2. The structure of a gas and its molecular weight are related. The structure or formula of a gas indicates the relative numbers and kinds of atoms that unite to form the gas molecule. For example, the formula CH_4 indicates that carbon and hydrogen are present in the compound in a 1 to 4 ratio. By taking the atomic weight of carbon and adding to it four times the atomic weight of hydrogen, the molecular weight of the gas CH_4 (methane) can be obtained. Table 2.2 shows the molecular weight of methane to be 16.043.

Pound-Atom and Pound-Mol

The mass in pounds of a given element, which is equal numerically to its atomic weight, is termed a pound-atom. Similarly, when the mass of a molecule is expressed in pounds numerically equal to its molecular weight, the term is referred to as a pound-mol. A pound-mol of methane (CH_4) would weigh 16.043 lb.

Mol Fraction

Another term that is quite useful in natural-gas operations is the mol fraction. If natural gas is composed of two hydrocarbon gases such as methane and ethane, there are two kinds of molecules. The number of methane molecules divided by the sum of methane and ethane molecules would represent the mol fraction of methane in the gaseous mixture. The mol fraction times 100 is the mol percent. For all practical purposes, the mol percent and volume percent are the same. This relationship holds only for gases and does not apply to liquid systems.

11

Table 2.3

VALUES OF R CORRESPONDING TO VARIOUS SYSTEMS OF UNITS

Units of Pressure	Units of Volume	R
Per gram-mol (temperature, degrees Kelvin)		
Atmospheres	Cubic centimeters	82.060
Per pound-mol (temperature, degrees Rankine)		
Pounds per square inch	Cubic inches	18,510.000
Pounds per square inch	Cubic feet	10.710
Atmospheres	Cubic feet	0.729

Critical Properties

Critical temperature, by definition, is the temperature above which a hydrocarbon gas cannot be liquefied no matter how much pressure is applied. For example, the critical temperature for methane is -116.6 F (table 2.1). If the temperature of methane gas is above this, it cannot be liquefied regardless of how much pressure is applied to it.

Table 2.2 shows another critical property of gases, critical pressure. The critical pressure is the pressure required to liquefy a gas at its critical temperature. In the previous example, methane has a critical temperature of -116.6 F. At this temperature, a pressure of 667.8 psia would be needed in order to liquefy the gas.

IDEAL GAS LAWS

From extensive experimental investigations, the ideal gas law has been empirically developed. In fact, the definition of the absolute scale of temperature is based on this relationship.

$$P_v = RT \qquad (1)$$

or

$$PV = nRT \qquad (2)$$

Where

R = a proportionality factor
T = absolute temperature
v = volume of one mol of gas
n = number of mols of gas
V = volume of n mols of gas
P = absolute pressure

When equation (1) is rearranged

$$R = \frac{P_v}{T} \qquad (3)$$

Assuming the validity of Avogadro's hypothesis that equimolar quantities of all gases occupy the same volume at the same conditions of temperature and pressure, it follows from equation (3) that the gas-law factor R is a universal constant. The Avogadro hypothesis and the ideal gas law have been experimentally shown to approach perfect validity for all gases under conditions of extreme rarefaction; that is, where the number of molecules per unit volume is very small. The constant R may be evaluated from a single measurement of the volume occupied by a known molar quantity of any gas at a known temperature and at a known *reduced* pressure.

In the use of the gas-law equations, great care must be exercised that consistent units are employed throughout. The pressure and temperature must be in absolute units such as psia and degrees Rankine. The numerical value of the gas constant R has been carefully determined and may be expressed in any desired energy units. (see table 2.3.)

When substances exist in the gaseous state, two general types of problems arise in determining the relationships among weight, pressure, temperature, and volume. The first type involves only the last three variables—pressure, temperature, and volume. For example, when given a specified volume at a specified temperature and pressure, the volume at any other specified pressure and temperature combination can be calculated. For such calculations, the weight of gas is not required. The second more general type of problem involves the weight of gas. For example, a specified weight of a substance exists in the gaseous state under conditions, two of which are specified and the third is to be calculated. Or conversely, it is desired to calculate the weight of a given quantity of gas existing at specified conditions of temperature, pressure, and volume. Problems of the first type may be readily solved by means of the proportionality indicated by the gas law. Equation (2) may be applied to n mols of gas at conditions P_1, V_1, T_1 and at conditions P_2, V_2, T_2.

$$P_1 V_1 = nRT_1$$
$$P_2 V_2 = nRT_2$$

Combining:

$$\frac{P_1 V_1}{P_2 V_2} = \frac{T_1}{T_2} \qquad (4)$$

This equation may be applied directly to any quantity of gas. If the three conditions of state one are known, any one of those of state two may be calculated to correspond to specified values of the other two. Any units of pressure volume and absolute temperature may be used, the only requirement being that the units in both initial and final states be the same.

The equation $PV = nRT$ is in a form to permit the direct solution of problems of the second type in which are involved both weights and volumes of gases. With weights expressed in molal units, the equation may be solved for any one of the four variables if the other three are known. However, the value of the constant R must be expressed in units to correspond to those used in expressing the four variable quantities.

The ideal or perfect gas law permits satisfactory calculations of pressure-volume-temperature relationships at low pressures where the volumes per mol are relatively large, the distance between molecules are great, and the temperatures are relatively high. However, under conditions of small molar volumes, corresponding to high pressures, the errors in assuming ideal gas behavior may be as great as 500 percent. Several hundred equations have been proposed by various scientists to express the correct PVT relationship of gases, but none has been found to be universally satisfactory. These equations are referred

to as the equations of state. All of these equations are cumbersome to use, and engineers have developed a substitute equation of state using a compressibility factor term that compensates for the deviations from the ideal gas law.

The equation of state may be written as follows:

$$PV = ZnRT$$

Z in this case is the compressibility factor, and it is a function of temperature, pressure, and nature of the gas. Compressibility factors for a large number of gases have been determined experimentally. They have also been tabulated in different ways over a wide range of temperatures and pressures. Usually, the compressibility factor is plotted or tabulated as a function of pressure and temperature for a given gas's specific gravity. The most useful relationship for compressibility is a plot of the compressibility factor as a function of reduced pressure and reduced temperature. In order to use a chart such as that shown in figure 2.3, it is necessary to know the critical temperature and critical pressure of the gas in question. The reduced properties of a gas are calculated as follows:

$$T_r = \frac{T}{T_c} \qquad\qquad P_r = \frac{P}{P_c}$$

Where the subscript r indicates the reduced and c the critical properties. Both temperature and pressure must be expressed in absolute units.

For gas mixtures such as those found in dealing with natural gas, critical properties are not easily found in books and tables; so general use is made of what is termed pseudocritical properties. The pseudocritical properties are calculated by using the mol fraction and the critical properties of each pure component. An example of this is shown in table 2.4. The gas used is a mixture of methane, ethane, and

Table 2.4
CALCULATION OF PSEUDOCRITICAL PROPERTIES

Component	Mol Fraction (N)	T_c (R)	NT_c	P_c (psia)	NP_c
Methane (CH_4)	0.015	343	5.1	668	10.1
Ethane (C_2H_6)	0.120	550	66.0	708	85.0
Propane (C_3H_8)	0.865	666	576.1	616	532.0
			$T'_c = 647.2$		$P'_c = 627.1$

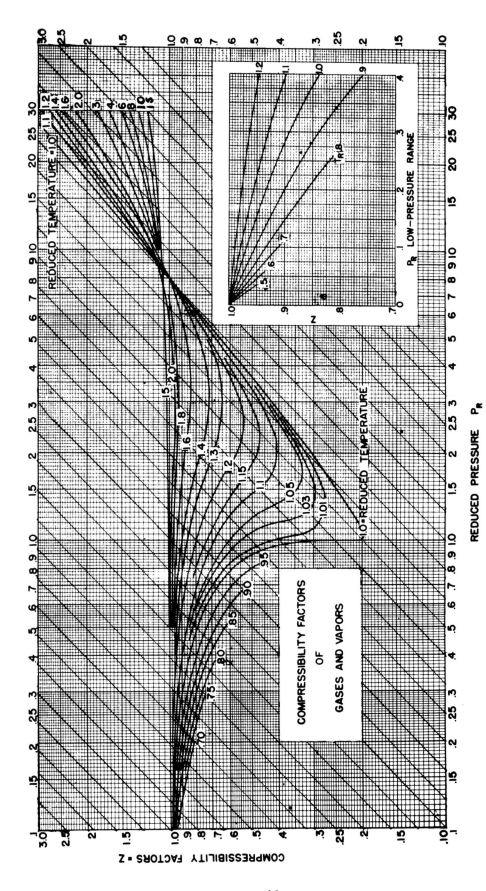

Figure 2.3. Compressibility Factors for Natural Gas

14

propane with the composition in mol fraction as given.

Using the pseudocritical properties calculated in the table and figure 2.3, the compressibility factor for this gas at 100 F and 300 psig can be determined as follows:

$$T = 100 \text{ F} + 460 = 560 \text{ R}$$

$$T_r = T/T_c' = \frac{560}{647.2} = 0.868$$

$$P = 300 \text{ psig} + 14.7 = 314.7 \text{ psia}$$

$$P_r = \frac{P'}{P'_c} = \frac{314.7}{627.1} = 0.502$$

Using $T_r = 0.868$ and $P_r = 0.502$, $Z = 0.64$.

Sometimes another term is used by engineers to compensate for the deviation from the ideal gas law. This term is referred to as the supercompressibility factor (F_{pv}) and is used primarily for high-pressure calculations. There is a mathematical relationship between supercompressibility and compressibility, as follows:

$$Z = \frac{1}{(F_{pv})^2}$$

The ideal gas-law equation substituting the supercompressibility factor for the compressibility factor can now be written—

$$PV = \frac{nRT}{(F_{pv})^2}$$

Tables of supercompressibilities for natural gas are available from a number of sources and can be readily used with the above equation to predict PVT relationships.

EQUILIBRIUM CONCEPTS

In handling natural gas mixtures, the engineer must be able to predict the exact relationship among all the components at any given temperature and pressure. Not only does he have to know whether he has a vapor, liquid, or a combination of both, but he also must be able to predict the distribution of each component in each phase. For this reason, engineers have devised the equilibrium constant or K value. The equilibrium constant is known as the vapor-liquid equilibrium ratio and is designated as Y_i/X_i, where Y_i is the mol fraction of component i in the vapor phase and X_i is the mol fraction of component i in the liquid phase. The K factor is a function of temperature, pressure, and the composition of a particular system. K values most widely used are those which have been developed by NGPA. In order to properly use the NGPA equilibrium constants, the engineer must be able to predict the convergence pressure of the natural gas stream being investigated.

Convergence pressure (P_k) is the pressure at a given temperature where the vapor-liquid equilibrium ratios for the various components in a system become or tend to become equal to unity (1.0). Figures 2.4, 2.5, and 2.6 are K charts for a convergence pressure of 2,000 psia for methane, ethane, and isobutane. Note that all the curves for various temperatures converge at a K value of 1.0 and a pressure of 2,000 psia.

There are several methods of determining the convergence pressure, but the most prevalent one is that based on the ideas of Hadden and Winn, which is discussed in the Natural Gas Processors Suppliers Association Engineering Data Book, 1972. It involves the selection of a pseudobinary system consisting of (1) the lightest component present in an amount of at least 0.1 mol fraction in the liquid phase and (2) the remaining components of the liquid that are heavier. The real problem is with the heavier components since selection must be made of one component from these that is representative of the group of components. A suitable method is to employ the weighted average critical pressure and temperature of the combined heavier components. The usual shortcut is to calculate the pseudocritical pressure of the heavier component portion and then pick the single component, such as butane, pentane, and so forth that has a critical temperature closest to the pseudocritical temperature of the heavier components. Using the lightest component and the component selected above as representative of the heavier components, a binary system is selected. Figure 2.7 is used to determine the approximate convergence pressure versus the system temperature. With this approximate value for the convergence pressure, the selection of the series of K values having a convergence pressure equal to or greater than the calculated convergence pressure is possible. Figure 2.7 is a plot of the convergence pressure for various binary systems as a function of temperature and is very useful to the

15

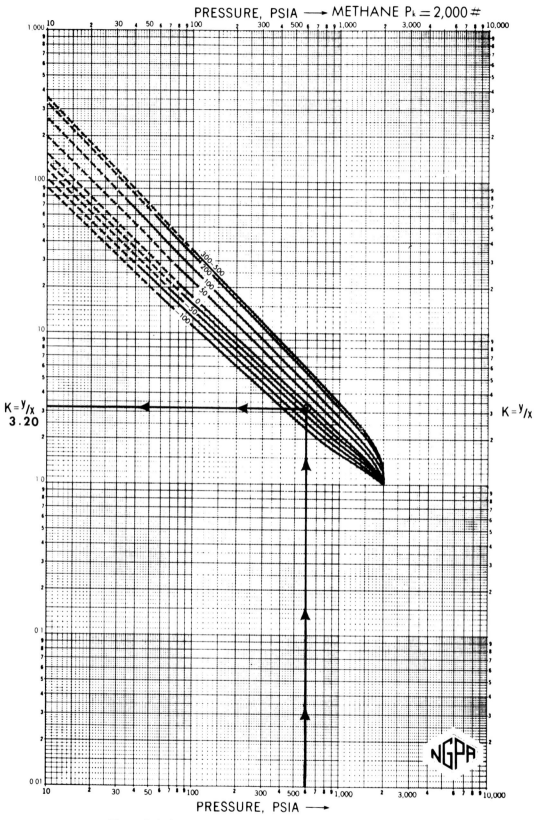

Figure 2.4. Convergence Pressures at 2,000 psia for Methane

16

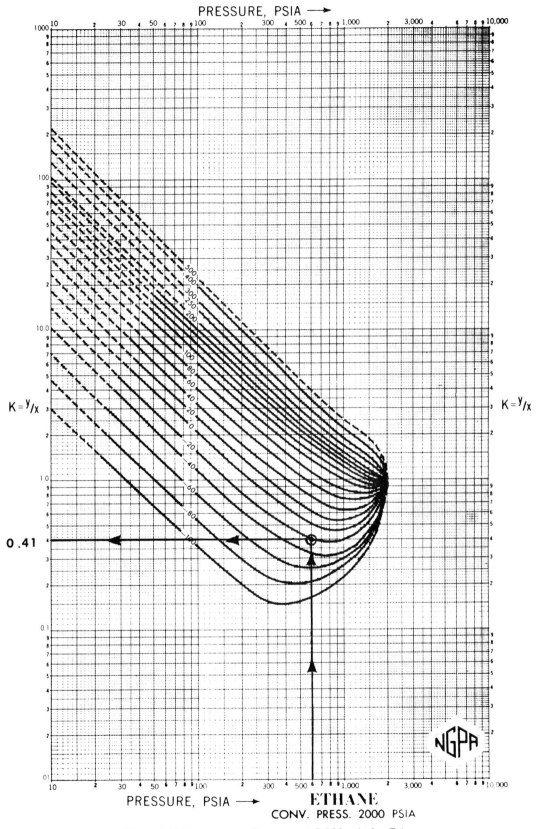

PRESSURE, PSIA →

K = y/x

0.41

PRESSURE, PSIA →   **ETHANE**
                                CONV. PRESS. 2000 PSIA

Figure 2.5. Convergence Pressures at 2,000 psia for Ethane

17

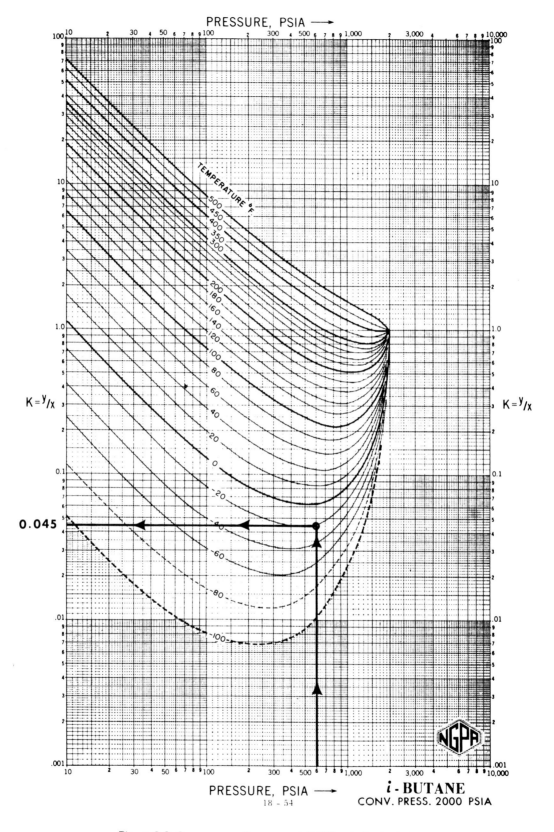

Figure 2.6. Convergence Pressures at 2,000 psia for Isobutane

18

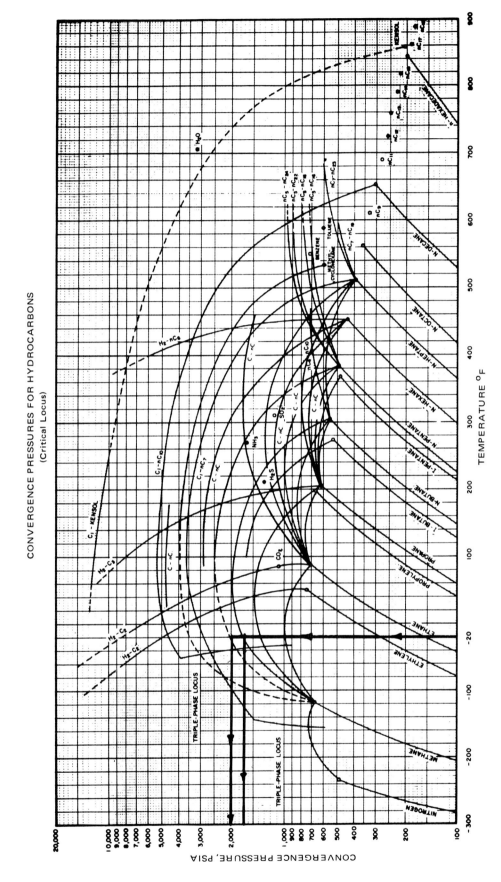

CONVERGENCE PRESSURES FOR HYDROCARBONS
(Critical Locus)

Figure 2.7. Plot of Critical Loci for Binary Systems—Convergence Pressure Estimates

19

engineer in determining the exact vapor-liquid relationship for any given system or any temperature and pressure.

In actual practice, an *assumed* convergence pressure is used to establish *estimated* K values and an equilibrium calculation performed to arrive at an *estimated* amount and composition of the liquid phase at given operating conditions for a given hydrocarbon system. This provides the basis for selecting the pseudobinary system referred to above. If the convergence pressure calculated from this binary system approximates the assumed convergence pressure, K values so established may be used with reliance in equilibrium calculations. If the assumed and calculated convergence pressures are far apart, the calculation must be repeated using an assumed pressure more closely approaching that obtained in the initial calculation.

K values have been established for a number of convergence pressures, specifically 800, 1,000, 1,500, 2,000, 3,000, 5,000, and 10,000 psia. To obtain the K value for a given component, it is only necessary to locate the intersection of the operating pressure and temperature curves on the appropriate convergence-pressure chart. For example, see figures 2.4, 2.5, and 2.6 where the K values for methane, ethane, and isobutane have been determined at 3.20, 0.41, and 0.045, respectively, for an operating pressure of 600 psia and temperature of −20 F on a 2,000 psia convergence pressure K chart.

Before proceeding with an illustrative problem of vaporization equilibrium calculations (flash calculation), two additional terms should be mentioned. These are *bubble point* and *dew point*. Bubble point is the temperature at which the first bubble would form in the liquid at a given pressure. In other words, it is the temperature at which the hydrocarbon mixture begins to vaporize at a given pressure. At the bubble point, the sum of the $K_i N_i$ = 1.0, where K_i is the equilibrium vaporization constant and N_i is the mol fraction for each component. Dew point is the temperature at which the first drop of liquid forms in the system at a given pressure. It is the temperature at which condensation takes place at a given pressure. At the dew point, the sum of N_i/K_i = 1.0.

It is wise for the engineer to check a system for its bubble point or dew point before proceeding with the vaporization equilibrium calculation as this will indicate whether or not there is a single-phase system (all liquid or all vapor) or a two-phase system (vapor and liquid). Flash calculations have no application

where a single-phase system exists.

The following problem has been taken from the Natural Gas Processors Suppliers Association *Engineering Data Book*, 1972 edition. It is included in order to show the general procedure followed in designing equipment to process natural gas streams and predicting the recovery of hydrocarbon liquids. The equilibrium constants (Ks) are those developed by the NGPA and published in the above reference. Only a few selected charts, figures 2.4, 2.5, and 2.6, are reproduced to show how the K values are determined for the example problem.

For reference, the variables and flash equations are defined as follows:

N_i = mol fraction of component i in the total mixture or system
X_i = mol fraction of component i in the liquid phase
Y_i = mol fraction of component i in the vapor phase
K_i = Y_i/X_i at thermodynamic equilibrium
L = mol fraction of liquid in the total mixture
V = mol fraction of vapor in the total mixture
Σ = summation of calculated terms indicated; that is, ΣX_i and ΣY_i

A situation of reproducible steady state conditions in a piece of equipment does not necessarily imply that classical thermodynamic equilibrium exists. If the steady composition differs from that for equilibrium, the reason can be the result of time-limited mass transfer and diffusion rates. This warning is made because it is not at all unusual for flow rates through equipment to be so high that equilibrium is not attained or even closely approached. In such cases, equilibrium flash calculations as described here fail to predict conditions in the system accurately, and the K values are suspect for this failure—when in fact they are not at fault.

Using the relationships $K = Y/X$ and $L + V = 1.0$ and by writing a material balance for each component in the liquid, vapor, and total mixture, one may derive the flash equation in various forms. A common one is

$$\Sigma X_i = \Sigma \frac{N_i}{L + VK_i} = 1.0 \qquad (1)$$

Other useful versions may be written as

$$L = \Sigma \frac{N_i}{1 + (V/L)K_i} \quad \text{or} \quad \Sigma Y_i = \Sigma \frac{K_i N_i}{L + VK_i} \qquad (2)$$

At the phase boundary conditions of the bubble point (L = 1.0) and the dew point (V = 1.0), these equations reduce to

$$\Sigma K_i N_i \ = \ 1.0 \text{ (bubble point)}$$
$$\Sigma N_i / K_i \ = \ 1.0 \text{ (dew point)}$$

These are often helpful for preliminary calculations where the phase condition of a system at a given pressure and temperature is in doubt. If the $\Sigma K_i N_i$ is less than 1.0, the system is all liquid and compressed above its bubble point. If $\Sigma N_i / K_i$ is less than 1.0, the sample is all vapor and may either be above its upper dew point or below its lower dew point. The $\Sigma K_i N_i$ and $\Sigma N_i / K_i$ cannot both be less than 1.0 at the same conditions

Problem

A typical high-pressure separator gas is used for feed to a natural-gas liquefaction plant, and a preliminary step in the process involves cooling to -20 F at 600 psia to knock out heavier hydrocarbons prior to cooling to lower temperatures where these components would freeze out as solids. What is the composition of the feed gas after the preliminary cooling?

Solution

The feed-gas composition is shown in table 2.5, column 1. P_k is first estimated to be 2,000 psia and Ks are obtained from 2,000 psia K charts and shown in column 2. Then the flash equation (1) is solved for three estimated values of L as shown in columns 3, 4, and 5. By interpolating, the correct value of L for ΣX_i = 1.0 is determined to be 0.037 as shown in columns 6 and 7. The gas composition is then calculated using $Y_i = K_i X_i$ as in column 8. This is the composition of the gas leaving the cooler.

To check the convergence pressure assumption,

the components heavier than methane in the liquid are converted to a weight-fraction basis as shown in columns 12 and 13 using the molecular weights in column 9. Using this composition (column 13) of the pseudoheavy component, the weighted average critical temperature and pressure is calculated using the T_{ci} and P_{ci} in columns 10 and 11; the results are shown at the bottom of these columns. By referring to table 2.2, the calculated T'_c of 348 F is seen to fall between n-butane and n-pentane and the calculated P'_c falls in line with the other paraffin hydrocarbons. Referring to figure 2.7, the critical pressure for the pseudobinary, or the convergence pressure, may be calculated at -20 F without actually sketching in a new locus:

$$P'_{kx} \ = \ \frac{(T'_{cx} - T'_{c4})}{(T'_{c5} - T'_{c4})} \ \text{X} \ (P_{k5} - P_{k4}) \ + \ P_{k4}$$

P_{k4} and P_{k5} are read from figure 2.7 at the points where the -20 F temperature line intersects the critical loci curves for methane n-butane (1,700 psia) and methane n-pentane (2,000 psia). T'_{cx} refers to the pseudobinary and comes from column 10, table 2.5 (807.8 R). T_{c4} and T_{c5} are critical temperatures of n-butane and n-pentane respectively expressed in degrees R and obtained from table 2.2.

$$P'_k \ = \ \frac{(807.8 - 765.6)}{(845.5 - 765.6)} \ \text{X} \ (2,000 - 1,700) \ + \ 1,700$$

$$P'_k \ = \ \frac{42.2}{79.9} \ \text{X} \ 300 \ + \ 1,700 \ = \ 1,858 \text{ psia}$$

This confirms the assumed P_k = 2,000 closely enough for this illustration. The more closely the operating pressure approaches convergence pressure, the more accurately the calculated P_k must check the assumed P_k because in this region the K values are most sensitive to the P_k value used.

Table 2.5
FLASH CALCULATION AT 600 PSIA AND -20 F

Column	1	2	3	4	5	6	7	8	9	10	11	12	13
				Trial values of L		Final L = .030							
Component	Feed gas composition N	$P_K = 2000$ K	$L = .02$ $\frac{N}{L+VK}$	$L = .04$ $\frac{N}{L+VK}$	$L = .06$ $\frac{N}{L+VK}$	$L + VK$	Liquid $X = \frac{N}{L+VK}$	Vapor Y	MW	T_c (°R)	P_c (psia)	gm	wt fr
C_1	.9010	3.2	.28549	.28952	.29368	3.13400	.28749	.91997					
CO_2	.0106	1.14	.00932	.00934	.00937	1.13580	.00933	.01064	44.01	547.8	1071.0	.4106	.0093
C_2	.0499	.41	.11830	.11508	.11203	.42770	.11667	.04783	30.07	550.3	709.8	3.5083	.0798
C_3	.0187	.097	.16252	.14047	.12369	.12409	.15070	.01462	44.09	666.3	617.4	6.6444	.1510
iC_4	.0065	.038	.11356	.08499	.06791	.06686	.09722	.00369	58.12	735.0	529.1	5.6504	.1284
nC_4	.0045	.024	.10340	.07138	.05451	.05328	.08446	.00203	58.12	765.6	550.7	4.9088	.1116
iC_5	.0017	.0088	.05939	.03509	.02490	.03854	.04411	.00039	72.15	829.0	483.0	3.1825	.0723
nC_5	.0019	.0062	.07286	.04135	.02886	.03601	.05276	.00033	72.15	845.5	489.5	3.8066	.0865
C_6	.0029	.0019	.13265	.06934	.04694		.09107	.00017	86.17	913.8	440.0	7.8475	.1784
$^*C_7+$	.0023	.00066	.11140	.05661	.03794	.03064	.07506	.00005	107.0	998.0	380.0	8.0314	.1826
		$\Sigma =$	1.16889	.91317	.79983		1.0089	.99972		807.8 −459.7 ——— 348.1	514.4	43.9905	1.0000

*Average -C_7+ - C_7 properties

22

NATURAL GAS PRODUCTION

Crude oil and natural gas are referred to collectively as petroleum. Crude oil is merely the heavier constituents that naturally occur in liquid form, while natural gas refers to the lighter constituents of petroleum that naturally occur in gaseous form. Thus, in the main, the geologic features that apply to the origin, migration, and accumulation of petroleum will apply equally well to crude oil and natural gas. This entails consideration of the natural features of the earth where commercial quantities of petroleum have been discovered all over the world.

PETROLEUM ACCUMULATIONS

For petroleum to accumulate, there must first be a source of oil and gas; second, a porous bed must exist, which is permeable enough to permit the oil and gas to flow through it—the reservoir rock; and third there must be a trap, which is a barrier to fluid flow so that accumulation can occur against it. Much knowledge has been obtained from experience and observations, but certain generalizations can be made.
   (1) Petroleum originates from organic matter.
   (2) To become commercial, the hydrocarbons must be concentrated.
   (3) Petroleum reservoirs are mostly in sedimentary rocks.

Migration of Petroleum

It is generally accepted that any present accumulation of oil and gas is a result of migration of widely dispersed and relatively small individual quantities of hydrocarbons to a more concentrated deposit as found in a reservoir. In some cases, the source material may be in close proximity to the present pool. However, it is believed that in most instances the organic source material from which petroleum was formed is widely disseminated in the sediments and that a present accumulation is the result of the combination of many minute portions of petroleum from near and far. Several natural forces and conditions that assist this migration include: (1) compaction of source beds by the weight of the overlying rock, thus providing a driving force tending to expel fluids through pore channels or fractures to regions of lower pressure and normally a shallower depth; (2) gravitational separation of gas, oil, and water in porous rocks that are usually water saturated; (3) pressure differential from any cause between two interconnected points in a permeable medium; and (4) faulting of the earth's strata.

Petroleum Reservoirs

The accumulation of oil and gas into a commercial deposit required a reservoir to contain the oil and gas along with some water and a trap, which represents a set of geologic conditions that retained the oil and gas in the reservoir until discovery.

A petroleum reservoir is a rock capable of containing oil, gas, or water. To be commercially productive, it must have sufficient thickness, areal extent, and pore space to contain an appreciable volume of hydrocarbons and it must yield the contained fluids at a satisfactory rate when penetrated by a well.

Sandstones and carbonates are the most common reservoir rocks. In order to contain fluids, the rocks must have porosity. The porosity may be classified as (1) primary, which represents the voids resulting from original deposition such as intergranular porosity of sandstone or (2) secondary, which resulted from later physical or chemical change such as dolomitization, solution channels, or fracturing.

Porosity is expressed as the ratio of void space to the bulk volume of the rock, usually expressed in percentage. Dependent upon the method of determination, porosity may represent either total or effective porosity. In many porous rocks, there are a certain number of blind or unconnected pores. Effective porosity refers to only those pores that are connected so as to permit fluid passage.

Permeability is a quantitative measure of the ease with which a porous rock will permit the passage of fluids through it under the pressure gradient. Like porosity, it is dependent upon rock grain shape, angularity, and size distribution. In addition, it is very strongly dependent on the size of the grains. The smaller the grains, the larger will be the surface area exposed to the flowing fluid. The additional drag or frictional resistance of the larger surface area lowers the flow rate at a given pressure differential, and thus the smaller grain size will result in a lower permeability.

Petroleum Traps

A *trap* is a set of geologic conditions that has stopped the migration of oil and gas and caused the oil and gas to be retained in a porous reservoir. Commonly these traps are related to structural highs (anticlines and domes) or against faults and unconformities. They may be placed in two general classes: (1) those in which the reservoir has an arched upper surface and (2) those in which there is an up-dip termination of the reservoir.

A simple form of trap is illustrated in figure 3.1. Figure 3.1 is a vertical cross section of a porous and permeable reservoir rock (such as a sandstone) that is overlain by a dense and impermeable bed (such as shale). It can be pictured that the oil and gas originated at a point located down-dip to the right or left of the fold. As gas and oil moved upward through pore passages of the water-filled reservoir rock, they encountered the sealing bed of shale or similar rock overlying the reservoir rock and continued to move

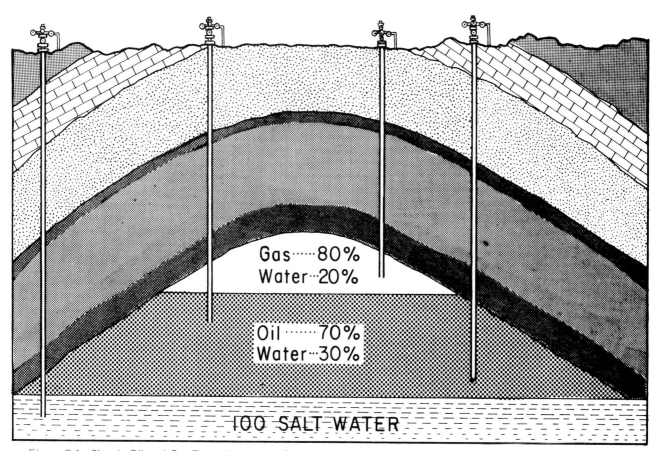

Figure 3.1. Simple Oil and Gas Trap. Percentage figures relate to the portion of the void space in the rocks occupied by fluids.

24

upward and laterally below the sealing surface until stopped by the attic of the fold. The gas, being lightest of the three fluids, would accumulate at the crest; and the oil, being next in density, would form a layer below the gas and above the water. Actually, all of the water originally contained in the pores of the reservoir rock would not be displaced by the accumulated oil and gas, and this would constitute the interstitial- or connate-water content of the reservoir. This nonreplaced water is of particular significance in making volumetric estimates of oil and gas reserves. Normally, for an intergranular type of reservoir, the interstitial-water content will be in the range of 20 to 50 percent of pore space and, in some cases, either above or below this range. The connate-water content is greatly influenced by the surface character of the sand grains, by the permeability of the sand, and by proximity to the level of 100 percent water. In general, oil-wet sands have a much lower connate-water saturation than water-wet sands. Also, sands of low permeability show higher connate-water saturations. Furthermore, in a reservoir with very small pores, there is a transition zone at the bottom where over a vertical interval the connate water gradually increases to 100 percent. This phenomenon explains why on the edge of many reservoirs significant percentages of oil saturation may be observed from cuttings and cores, yet the production will be all water. Transition zones are more extensive and of more significance in oil reservoirs than in nonassociated gas reservoirs.

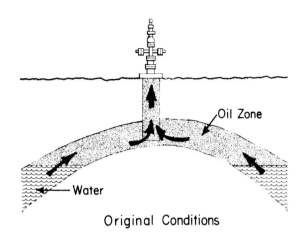

Original Conditions

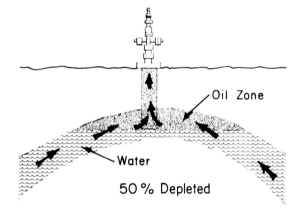

50% Depleted

Figure 3.2. Water-Drive Reservoir

Associated and Nonassociated Gas

Natural gas produced from a reservoir that contains oil is called *associated gas.* The term applies to both free gas from a gas cap and to solution gas. In general, the term casing-head gas is synonymous with associated gas since it commonly refers to gas production from oil wells. Gas produced from a reservoir that does not contain oil is referred to as *nonassociated gas* on the basis that it is not directly associated with oil underground. In certain fields, these terms assume particular importance because regulatory and control measures applied to associated gas are directed toward a consideration of the effect of gas production upon oil production rates and ultimate recovery, whereas the production of nonassociated gas does not involve such considerations.

Reservoir Drives

There are two general types of reservoir drives: *depletion drive* and *water drive.* Depletion drive is operative where the reservoir can be termed a closed reservoir; that is, the hydrocarbon accumulation is not in contact with a large body of permeable water-bearing sand. Expansion of the hydrocarbons and other reservoir materials that occurs as pressure is reduced furnishes the only energy for movement of fluids through the formation and to the surface. With regard to oil reservoirs, there may be two different types of depletion drive: solution-gas drive and gas-cap drive.

Figure 3.2 represents a water-drive reservoir. With the oil zone as shown, this would be an oil reservoir without a gas cap. Similarly, it could be a gas reservoir with gas instead of oil accumulated in the crest of the structure. In either case, the water is free to move into the reservoir and displace the hydro-carbons as they are withdrawn. Thus, the reservoir pressure remains essentially constant, and the volume occupied by remaining recoverable hydrocarbons becomes less and less as production proceeds. Because of the greater displacement efficiency of water as compared to gas, the oil recovery from a water-drive reservoir is usually much higher than for either of the depletion drives. Recovery from a good active water drive may be in the order of 75 percent of the oil in place, depending on rock characteristics.

Producing Gas Wells

So far in this discussion, most of the features relating to origin, migration, accumulation, reservoir rock and fluid characteristics, and types of reservoir drives have been discussed as applying to both gas and oil. Even though this discussion relates principally to natural gas, it has been necessary to consider crude oil. Because of the close interrelationship of gas and oil in nature and because a very significant part of the present total natural gas production is associated gas produced with oil, natural gas production is influenced in numerous ways by oil development and production practices. However, there are certain additional features relating more exclusively to gas production that should be presented.

Reservoir pressure is a controlling factor in the ability of a well to produce. A decrease in the bottom-hole pressure of a gas well is reflected by a drop in its productivity. Reservoir pressure is always a consideration in the final stages of depletion of a gas reservoir. For most gas reservoirs, the estimated reserves, the forecast of producing rates, and the estimated producing life are based on a prediction of pressure decline and estimated abandonment pressure. Considerations also usually include a determination of the economic feasibility of gas compression at the wellhead to boost the pressure of the gas for sale into a pipeline and lower the abandonment reservoir pressure.

For a given production rate, pressure will decline more rapidly in a small reservoir than in a large reservoir, if both are of depletion type. Water influx into a gas reservoir can help support reservoir pressure, and this may tend to mask the difference between a small and a large reservoir since either may have an active water drive sufficient to support reservoir pressure at the prevailing rate of production. It is possible to misinterpret a low rate of pressure decline as being due to a high gas reserve when actually the reservoir pressure is being at least

partially maintained by the influx of water causing a gradual reduction in reservoir size, and there is a remaining gas reserve much less than that estimated on the premise of a depletion type of drive. Thus, any other clue that will better support a conclusion regarding the type of drive will be helpful in reaching a more reliable reserve estimate. In a multiwell gas field, a possible clue regarding the type of drive may be obtained by a study of the water production trends on different wells having various elevational levels of completion in the reservoir. An extremely rapid bottom-hole pressure drop may be caused by low sand permeability either throughout the reservoir or immediately surrounding the wellbore.

Estimation of Reserves

In order to understand the art and science of reserve estimation, it is necessary to define certain terms that are commonly used and recognize the sources of the data represented by these terms. Some of these terms have already been discussed in chapter two. Others, referring particularly to reservoir factors, are as follows:

Abandonment pressure is the average reservoir pressure at which insufficient gas is expelled to permit continued economic operation of a producing gas well. The value will vary from a few pounds per square inch (psig) up to 500 psig or even 1,000 psig depending on purchaser line pressure and the amount of compression that can economically be installed.

Condensate liquids are hydrocarbons that are gaseous in the reservoir and will separate out in liquid form at the pressures and temperatures at which separators normally operate; sometimes they are called distillate. The gravity of a condensate or distillate, is high—usually above 50° API.

Condensate ratio is the ratio of the volume of liquid produced to the volume of residue gas produced and is usually expressed in barrels per million cubic feet (bbl/MMcf).

Gas saturation is the percentage of the total pore space occupied by gas. Within the pore structure of the reservoir rock, a portion of the pore space may be occupied by gas.

Hydrocarbon pore volume is the volume of the pore space in the reservoir occupied by oil, natural gas, or other hydrocarbons (including nonhydro-carbon impurities) and may be expressed in acre-feet, barrels, or cubic feet as appropriate. An acre-foot is the volume equivalent to an area of one acre, one foot thick.

Pore volume is the volume within the reservoir (in acre-feet, barrels, or cubic feet) not occupied by rock.

Porosity is the percentage of the total reservoir that is not occupied by rock. Various methods to determine porosity are available including electric logs and core analysis.

Permeability is the term used to describe the flow capacity of a reservoir rock. Permeability may be reported in darcys or millidarcys. Well-log correlations, core analysis, and pressure buildup (or drawdown) tests are the usual sources of permeability data. Absolute permeability is the permeability of a rock to a single fluid if the rock is 100 percent saturated with that fluid. Effective permeability is the permeability of a rock to a fluid when the saturation of that fluid is less than 100 percent.

Pressure depletion is the method of production of a gas reservoir that is not associated with a water drive. This is the process where gas is removed and reservoir pressure declines until all the recoverable gas has been expelled.

Reservoir pressure is the average pressure within the reservoir at any given time. Determination of this value is best made by bottom-hole pressure measurements with adequate shut-in time. If a shut-in period long enough for the reservoir pressure to stabilize is impractical, then various techniques of analysis by pressure buildup or drawdown tests are available to determine static reservoir pressure.

Reservoir temperature is the average temperature within the reservoir and is measured during logging, drill-stem testing, or bottom-hole pressure testing using a bottom-hole temperature recorder.

Retrograde reservoir is a reservoir where the pressure is high and the hydrocarbon content is completely single phase at initial conditions, and as the pressure drops due to production, the heavier hydrocarbon components condense forming liquids within the reservoir. This action is the retrograde phenomenon. If the reservoir pressure is completely depleted, only a small portion of these liquids will revaporize and be recovered.

Residual gas saturation is the portion of hydrocarbons that cannot be removed by ordinary producing mechanisms when a porous reservoir has been saturated with hydrocarbons. This value is usually a specific percentage of the pore volume. In the case of gas, the volume measured at standard conditions that is retained in a reservoir as residual gas saturation is an inverse function of the pressure due to the effect of the gas laws.

Solution gas is gas dissolved in the oil. Except for possibly some very heavy oil or tar reservoirs, all oil reservoirs contain some solution gas. Solution gas may be compared to the carbon dioxide in a soft drink. If more gas is present than can be dissolved at reservoir conditions, a free gas phase will exist and a gas cap may form. Under these conditions, the oil reservoir is said to be saturated. If less gas is present than can be held in solution, the reservoir is referred to as undersaturated.

Solution-gas drive is the mechanism where the solution gas is the principal source of energy to expel the oil from the reservoir.

Water saturation (S_w) is the percentage of total pore space occupied by water. Virtually all reservoirs contain some water. In some cases it may be the irreducible minimum or connate-water saturation. Others may have a mobile water phase present. In all cases, water reduces the space available for hydrocarbons by its presence. Water saturation is normally estimated by electric log calculations although cores specially obtained may be analyzed for this value.

The initial estimate made on a newly discovered gas reservoir will usually be a volumetric calculation. Factors needed to arrive at this estimate would include the type of reservoir, its area, its thickness, its porosity, and its water saturation, and the temperature, pressure, and composition of the gas. While in practice several of the following steps may be combined into a single calculation, each step will be separately computed here for purposes of illustration.

1. Determine the hydrocarbon pore volume per acre-foot of reservoir.

$$43,560 \times \text{Porosity} \times \frac{100 - \% \text{ Water Saturation}}{100}$$

$$= \text{Cubic Feet per Acre-Foot.}$$

2. Determine the volume of gas at standard conditions initially contained in an acre-foot of reservoir.

Gas Volume (cu ft) = Hydrocarbon Pore Space

per Acre-Foot X $\dfrac{\text{Original Reservoir Pressure (psia)}}{14.7 \text{ (psia)}}$

X $\dfrac{520 \text{ R}}{\text{Reservoir Temperature R}}$

X $\dfrac{Z \text{ Factor (14.7 psia)}}{Z \text{ Factor (Original Reservoir Pressure)}}$

3. Determine the volume of gas at standard conditions contained in an acre-foot of the reservoir at estimated abandonment pressure.

Gas Volume (cu ft) = Hydrocarbon Pore Space

per Acre-Foot X $\dfrac{\text{Abandonment Pressure (psia)}}{14.7 \text{ (psia)}}$

X $\dfrac{520 \text{ R}}{\text{Reservoir Temperature R}}$

X $\dfrac{Z \text{ (14.7 psia)}}{Z \text{ Reservoir Abandonment Pressure (psia)}}$

4. Subtract volume (3) from volume (2) to obtain recoverable gas reserves per acre-foot of reservoir.

5. From combined geological and engineering studies, the wells' drainage area and pay thickness are determined and the effective reservoir volume in acre-feet established. The effective reservoir volume multiplied by the recoverable gas per acre-foot results in the total recoverable gas available.

6. If it is known that an active water drive is present, the ultimate recovery is usually estimated by applying a recovery factor to the original volume in place. The recovery factor is determined by laboratory analyses on cores taken from the reservoir rock.

After production begins, it is necessary to observe the performance of the reservoir as it is depleted to determine if the initial estimate is reasonable or should be adjusted. The typical method of evaluating performance is a plot of bottom-hole pressure over the appropriate deviation factor at the time for the

pressure involved (P/Z) versus the cumulative recovery (fig. 3.3). Once sufficient history has been obtained under reasonably stabilized operation conditions, it is possible to extrapolate the historical plot to the anticipated abandonment pressure and thus arrive at an estimate of ultimate reserves.

Unfortunately, several factors will affect the validity of this method of estimation. If a full or partial water drive is present, the rate of pressure decline will be less than would have been observed had the reservoir been on a straight pressure depletion as shown in figure 3.3. Such a decline would be erroneously interpreted as indicating a much larger reservoir than actually exists. The performance observed must thus be closely tied to the known

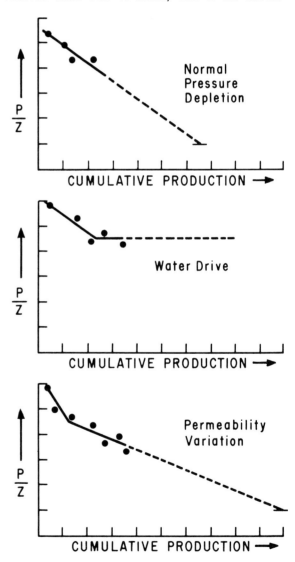

Figure 3.3. Effect of Reservoir Characteristics on Production Performance

28

geological configuration of the reservoir so that the reasons for the particular performance observed can be properly evaluated.

In certain other cases, it may be found that the gas reservoir is actually a gas cap on the top of an oil reservoir. In this case, the pressure decline will be influenced by the rate of oil production and the evolution of solution gas from the oil into the gas cap as the pressure declines. In such a case, the entire reservoir must be analyzed including the effects of the oil production rate and the effect of pressure reduction on solution gas. Transfer between the oil and gas cap can go either way depending on reservoir conditions of temperature, pressure, and fluid composition.

Variation in the reservoir permeability can also affect the observed pressure performance. A typical pattern is a sharp drop of the reservoir pressure during the early depletion—which will yield a very low ultimate recovery if extrapolated in a straight-line manner—but which breaks to a lower slope on a balance between the flow capacity of the unfractured matrix and the fracture system that has been established, as shown in the lowest diagram of figure 3.3. In such cases, it is necessary to pass this breakpoint before any kind of reliable pressure decline estimate can be made.

In order to determine reserves by the pressure decline versus cumulative method, ideally one should have accurate bottom-hole pressure data based on a shut-in period that is sufficient for the well to achieve static conditions and accurate cumulative production histories. Although current practices normally yield adequate production data, bottom-hole pressure data is often sparce and even when available is based on short shut-in periods. The most usual test provided is a shut-in wellhead pressure test that was obtained to meet regulatory requirements or as a result of the well having been shut in for some other purpose.

In the event that the well produces a dry gas without condensate or water present, it is a very simple matter to arrive at a reasonable estimate of bottom-hole pressure from a shut-in wellhead pressure valve. If the well is producing condensate and/or water with the gas stream when it is shut in, there will usually be an accumulation of liquids in the bottom portion of the wellbore that will account for a substantial part of the bottom-hole pressure. Unless the volume of the liquid is reasonably known, it is impossible to estimate the bottom-hole pressure from the shut-in surface pressure. In such situations, bottom-hole pressure recordings are imperative.

Gas-Well Rating

The performance of a natural gas well depends upon the physical properties of the reservoir rock, the extent and geometry of the drainage area, the properties of the flowing fluid, and the conditions of pressure distribution within the drainage matrix. The relation between the daily rate of delivery of the gas and the pressure drop within the reservoir is characteristic of the behavior of each well and is usually referred to as the back-pressure behavior of the well. A test made to determine quantitatively this relationship for a given well is referred to as a back-pressure test.

A back-pressure test consists of determining the static formation pressure (P_s) when the well is shut in and the flowing bottom-hole pressure (P_f) at each of two or more flow rates (Q). A value of $P_s^2 - P_f^2$ is then calculated for each flow rate and plotted on log-log paper against the corresponding value of Q. A plotting of this type is illustrated by figure 3.4. It has been well established that for normal, single-phase gas wells this plotted relationship will be a straight line, the general equation of which can be written

$$Q = C (P_s^2 - P_f^2)^n$$

Where C is the performance coefficient, and n is the exponent corresponding to the slope of the straight line plotted on log-log paper. The determination of

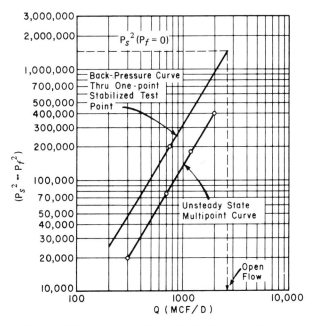

Figure 3.4. Isochronal-Type Back-Pressure Test for Gas Wells

this relationship for a given well permits the analysis of many operating problems and also provides a basis for predicting future behavior of a well or group of wells. For example, once correct stabilized values of C and n are established for a given well, the equation may be used as a basis for calculating (1) the effect of different sizes of producing string on the gas flow rate attainable from a well; (2) the effect of surface back pressure on flow rate or, in other words, the gas deliverability of the well against any chosen back pressure, such as gas-line pressure at the surface; (3) flow rate or deliverability of a well to be expected at some future time when the static pressure has declined to some chosen value; and (4) maximum flow rate, normally termed *absolute open flow,* that a well could theoretically deliver with zero pressure at the sand face. Although a back-pressure test of this type is relatively simple and the resultant C and n values are usable directly for a well that stabilizes quickly, the test has been found to require special treatment when dealing with tight wells; that is, wells that produce from reservoirs of low permeability and thus stabilize slowly. Such wells show slow buildup of bottom-hole pressure on a buildup test, and the flowing pressure does not stabilize quickly at a constant value on a flow test.

For slow-stabilization wells, a method called isochronal testing has been developed whereby a true value of n can be determined. With this method, the well is opened at one restricted rate of flow and pressure data are taken at specific time intervals without disturbing the rate of flow. The well is then shut in and allowed to return to a shut-in condition comparable to that existing at the time the well was first opened. It is then opened at a different rate of flow with data being taken at the same time intervals as before. This procedure is repeated as many times as necessary to obtain the desired number of data points. Extensive testing has proved that points taken at comparable time periods at different flow rates will yield a true value of n, whereas true values of n would not be determined if the points were taken randomly with regard to time as on a conventional back-pressure test. A constant n value was obtained by using only the points read at one time point when using the various flow rates. From a practical standpoint, except for check purposes, it would not be necessary to take more than one time reading at each flow rate in order to determine n.

In general, the performance coefficient C is a function of (1) effective permeability; (2) formation thickness; (3) gas viscosity; (4) gas gravity; (5) gas compressibility factor; (6) temperature of formation; (7) radius of drainage; (8) wellbore radius; and (9) a factor v, which varies with flow rate Q and slope n.

The factors C and n are considered to be constant throughout the life of a well, and this is essentially true except for certain things that sometimes happen to any producing well. For example, gradual plugging of porosity near a well by scale or wax deposits might materially change its flow characteristics. Also, condensation of water or liquid hydrocarbons near a wellbore might have a similar effect. Certain stimulation treatments such as fracturing or acidizing may often change the flow characteristics of a well, and a retest would be required to determine applicable C and n values.

Open-flow and deliverability tests of gas wells are often required by regulatory agencies, and specifications for such tests are usually stipulated. Gas-sales contracts may provide for determination of gas deliverability of each connected well. Allowables assigned by regulatory bodies and rates of gas taken by gas purchasers are often apportioned on the basis of open-flow capacity or deliverability as determined by these tests.

Accuracy of a test may be affected by several factors, which include—

1. skill and experience of the tester;
2. proper testing equipment;
3. wellbore conditions, particularly presence or absence of liquids;
4. accuracy of pressure determinations; and
5. high content of nonhydrocarbon material (H_2S, N_2, and CO_2) in the gas.

WELL EQUIPMENT

Gas-well producing problems literally start at the bottom of the well. The selection and size of casing, tubing, and wellhead equipment must take into account the expected rates of flow, fluid erosion, and chemical corrosion. Casing size will limit the size tubing that can be installed, and tubing size will limit the gas flow that can be produced with a given pressure drawdown. Safety shutoff equipment, some of which functions by flow velocity, is mandatory by government regulations in many cases and desirable in most gas wells. This equipment is usually installed in the tubing and, in many cases, at the wellhead. High rates of flow can cause problems of erosion in the tubing and at the wellhead. Corrosion due to chemical action can be serious in gas wells, and its effects must be taken into account at all points in the flow system. Deep wells usually have more serious corrosion problems than shallow wells because the flowing temperatures are higher. Deep well gas sometimes contains organic acids and hydrogen sulfide as well as carbon dioxide that will cause serious corrosion problems if protective measures are not taken. Some of the means to mitigate gas-well corrosion have been special alloy steels, plastic coatings, chemical injection, and displacement of chemicals into the well to obtain a protective coating on the tubing and other parts of the well exposed to corrosive fluids.

Tubing and Packer

Gas-well producing capacity depends upon the reservoir characteristics, bottom-hole pressure, restrictions of the tubing string, and pressure required at the wellhead for flow into a pipeline. If the producing pressure at the surface is less than pipeline pressure, compression will be needed to produce the well. Producing rates may vary widely, depending upon well capacity, the gas-sales contract, and the daily take of the transmission line. Production from a low-pressure, shallow well usually will not vary too much because of the narrow margin of available pressure. A deep gas well with high formation pressure may be required to produce 500 Mcf one day and 5,000 Mcf or more the next. Varying producing rates can cause wide temperature variations, with varying stress on the tubing string because of expansion or contraction as the temperature of the flow stream changes. The tubing and packer arrangement in the well must be able to handle temperature variations without damaging the integrity of the tubing string.

The final string of pipe usually run in a well is the tubing. The small diameter of the tubing permits more efficient production of fluid than casing and makes possible a safer completion because fluid may be circulated down the tubing and up the casing to remove undesirable fluid in the well. Tubing constitutes a string of pipe that can be removed if it becomes plugged or damaged. Tubing, in conjunction with a packer, keeps well fluid and pressure away from the casing because the packer seals the annular space between the tubing and casing.

Plastic linings have been used in the past to minimize the damage effect of corrosive compounds in the well stream on the tubing. Experience has shown that it is most difficult to obtain complete protection with plastic coatings; frequently the places left uncoated become "hot spots" where corrosion proceeds at accelerated rates. Wire line and other tools run into the tubing string will damage the plastic coating. Some operators use uncoated API J grade tubing and rely upon chemical inhibitors to minimize the damaging effects of acid gas (H_2S and CO_2) corrosion.

Corrosion Inhibitors

Gas-well corrosion is generally handled by the injection of chemical inhibitors to counter the chemical reaction between the acid solutions in the gas and the iron of the tubing and so forth. This is usually accomplished at the wellhead, either by batch treatments or continuous injection.

Weighted sticks of soluble material that were dropped into the tubing were popular at one time, but this practice has now been practically abandoned. Injection down the annulus between the casing and tubing where this is feasible or using small diameter tubing—kill string—inside the production tubing is sometimes used. A kill string reduces the capacity of the tubing for gas flow and cannot be used below about 12,000 ft because small diameter pipe will pull apart due to its own weight.

A common procedure for downhole chemical treating is injection by batch displacement down the tubing. This process enables an even coating on the inside of the tubing for physical protection from acid reactions as well as chemical inhibition. Squeeze

Figure 3.5. Positive and Adjustable Chokes on a Triple-Zone Completion

31

injection of chemical solution batches through the perforations into the producing interval will permit longer periods of time between treatments than batch displacement, but formation damage can take place and well productive capacity may be reduced.

Wellhead Equipment

Valves and fittings are installed at the wellhead to enable shutoff and regulated flow from the well. If corrosive conditions are expected, this equipment will be fitted with stainless trim to extend its service life. Figure 3.5 illustrates a high-pressure wellhead. Flanged or studded fittings have been standardized by the American Petroleum Institute (API) following sizes and ratings established earlier by American National Standards Institute (ANSI). (See table 3.1.)

Table 3.1

SIZE AND PRESSURE RATING OF FITTINGS

Designation		Pressure Rating, psi, 100 F	
ANSI	API	ANSI	API
600	2,000	1,440	2,000
900	3,000	2,160	3,000
1,500	5,000	3,600	5,000
- -	10,000	- -	10,000
- -	15,000	- -	15,000

A *choke* is a restriction used in the assembly of valves and fittings (Christmas tree) for the purpose of regulating flow. Chokes may be positive or adjustable as shown in figure 3.5.

Flow regulators are sometimes employed as wellhead equipment for a gas well to obtain rates of flow according to programmed requirements for a computerized flow controller. These are normally motor-operated diaphragm valves as described in a following chapter.

Safety valves are frequently installed as wellhead equipment to obtain automatic shutoff in the case of overpressure or underpressure conditions downstream of the well (fig. 3.6). This protection will operate in the case of a flow line plugged by hydrates or a ruptured flow line, which might cause a blowout. Governmental regulations require automatic shutdown equipment on offshore wells for protection in case of fire or storm damage. This equipment is also used downhole in the tubing string.

Corrosion detection devices are frequently installed as wellhead equipment in order to obtain

Figure 3.6. Wellhead Safety Shutdown Valve

information as to the corrosion that may be taking place in a well. These devices include containers to accumulate water samples on which iron-count observations can be made and the placement of various types of metal strips (coupons) to permit assessment of corrosive chemicals in the well stream.

GATHERING SYSTEMS

The facility with which gas is collected from the wells, controlled and conditioned when required, and transported to its primary destination is called a gathering system. Each gathering system is unique; it must be designed and installed to function safely, reliably, and economically within the environment of a particular field and set of physical and economic conditions. Gathering systems normally fall into one of four categories: (1) the single-trunk system with laterals; (2) the loop system, in which the main line is in the shape of a loop around the field; (3) the multiple-trunk system, in which there are several main lines extending from a central point; and (4) combinations of the above. Selection of the most desirable layout requires an economic study of several possibilities and depends on many variables, such as the type of reservoir, the shape of the reservoir, the way in which the land over the reservoir is being used, the quantity of impurities in the gas, the available and permissible flow rate of gas and liquid, the flowing and shut-in pressure and temperature, the climate and topography of the location, and the primary destination of the gas. Usually a separate economic study will be needed to justify the installation of automated or remote-control facilities.

Type of Reservoir

Petroleum reservoirs vary from those with undersaturated crude oil, like the East Texas field, to those with gas but practically no recoverable liquids, like the San Juan Basin fields in New Mexico. Solution gas and gas from associated reservoirs tend to be rich with the heavier hydrocarbon components that liquefy easily and must often be separated before the gas can be handled by the gathering system. Gas from some nonassociated reservoirs is produced with a proportion of liquid to gas low enough to permit the liquid to be gathered in the same pipe with the gas to a central point.

The shape of the reservoir, the configuration of the leases, and particularly the surface location of the wells and tanks served by a gathering system will have an influence on the system layout. For example, long narrow fields are usually best served by main-trunk systems. Likewise, a group of leases around the outer perimeter of a field would likely require a loop system.

Surface Usage

How the land over a reservoir is used has a most important bearing on the type of gathering system employed. If the land is used for agriculture, the surface owner will likely insist on lines and other facilities being located so as to interfere as little as possible with his farming operations, that is, planting, irrigation, harvesting, and so forth. At some locations, noisy equipment such as compressors, pumps, treaters, and heaters will need special consideration as to location and noise control. Special emphasis may be placed on safety devices and equipment. Gathering systems for reservoirs located beneath bodies of water are also unique. The design of such systems must take into account assessibility and protection of the

environment. Again special safety precautions must be provided.

Impurities in the Gas

Impurities are common in the production of natural gas. Some of these, such as sand and excessive amounts of water, must be removed immediately so that the gas can be gathered. Others, such as carbon dioxide, hydrogen sulfide, nitrogen, and small quantities of water, sometimes may be allowed to remain in the gas until it is gathered to a central location where the impurities can be removed more economically in a single larger plant than would be the case using small units located at each well.

Gas Flow Rate and Quantity of Liquids

The flow rate of associated gas is determined by the rate and method of producing the oil in the field. The flow rate of nonassociated gas production is limited by the deliverability of the well, governmental regulations, economic considerations, and contractual obligations. As the gas-flow rates become larger, larger pipe and other equipment with larger capacities will be required. If large quantities of liquid are allowed to flow with the gas in the pipe, flow becomes erratic, the liquid moves in slugs, and the gas-flow efficiency decreases. This is particularly serious in hilly country. To prevent or minimize this problem, scraping or slug-catching equipment is installed or facilities are installed for handling the liquids separately.

Pressure and Temperature

Gas produced at high pressure has more value because the potential energy expressed as pressure can be used to move the gas to another location for processing or use; of course, it also requires stronger piping and valves to control the gas. Sometimes pressure in excess of that required to move the gas is expended by using a pressure-reducing or throttling valve so that lighter pipe may be used for the flow system. If this is done, it may be necessary to install heaters to prevent the formation of hydrates in the gas, which has become cold due to the refrigeration effect of the rapid drop in pressure. Temperature of the flowing gas affects the design of the system, too. High temperature necessitates installation of equipment and features to control movement of the pipe caused by expansion of the hot steel. On the other hand, low gas temperatures present the potential problem of hydrate formation in the gas and consequent flow stoppage. To minimize the hydrate problem, gathering lines are installed without sharp bends where hydrates might accumulate and lines are often buried deeper to minimize heat loss during cold weather. Other and more costly methods are the addition of heat by means of heaters, the injection of hydrate-inhibiting chemicals such as methanol or glycol, and the removal of water vapor by the use of dehydration equipment.

Climate and Topography

Cold climates present continual or at least seasonal difficulties in relation to hydrates, equipment operation, and worker exposure. The climate must be considered when designing the gathering system. Location of the system in relation to topographical features such as mountains, swamps, open water, rocky terrain, and remote locations also affect the system design.

PRODUCING EQUIPMENT

The gathering system facilities can be grouped in two major categories. First, there are those items of basic equipment that are needed to transport the gas; to start, stop, and measure the flow; and to control and measure gas pressure and temperature. Secondly, there is the equipment needed to condition the gas so that it will flow safely and reliably.

Natural gas has two primary commercial uses: (1) gaseous fuel and (2) chemical feedstock. Sometimes a portion of the produced gas is returned to the reservoir to increase total hydrocarbon recovery. It is often economically attractive to process gas in order to separate the lightest component, methane, for use as gaseous fuel from the heavier components, which may find more valuable uses as liquid fuels or chemical feedstocks. Regardless of its ultimate use, gas usually will need to be conditioned in one or more ways in order for it to flow safely and reliably from the well to a process plant or transmission pipeline. The gas may require liquid and sand removal by mechanical separation, water-vapor removal by dehydration, an increase of temperature by heaters, a drop in pressure by chokes, or an increase in pressure by compressors.

Basic Equipment

The pipe through which the gas flows is the basic and most important part of the gathering system. It is made of a steel, which is selected in accordance with proper codes and regulations, and is usually welded together. Older pipelines and some small lines today are coupled together with threaded joints. The diameter of the pipe is determined by the amount of gas that is to be transported through the pipe, and the thickness of the pipe wall is determined by the pressure of the gas.

Flow of the gas is stopped or allowed to continue by the use of valves. Block valves are used to stop the flow; they are usually completely open or completely closed. Throttling valves to regulate the rate of flow are usually only partially open or closed; the inner valve is often positioned by a mechanical operating device controlled automatically by instruments or sometimes by a remote-control arrangement.

Various instruments are needed to measure the condition of the gas. Some of these instruments are merely indicating devices that can be useful to the operator at the job site, such as liquid-level gauge glasses, mercury thermometers, and dial-type pressure gauges. Others record particular information, such as a flow recorder, which measures the flow rate and pressure and then records that information on a paper chart. Still others use the measured information to adjust the position of a control device, such as a temperature controller adjusting the gas burner control on a heater or a pressure controller changing the position of the inner valve of a pressure-reducing valve. Safety devices are also within the general description of instruments and controls; however, their function is to prevent conditions from exceeding safe limits rather than reducing them for operational purposes. Various governmental regulations deal with the safety of flow systems and equipment.

Some degree of automation has been used in gas-gathering systems for years, such as the control of pressure, flow rates, and temperatures. Expansion of the use of such regulating equipment to permit automatic control of whole systems is now common practice. The use of electronic communications and computers often enables field operations of a very complex nature to be directed instantly from a remote location accurately and automatically. Such systems are expensive and have unique maintenance problems. On the other hand, these systems can be operated with a very small work force and can quickly, often automatically, adjust the system to wide ranges of demand. They usually provide the capability of rapidly determining the location of operating difficulties.

Gas-Conditioning Equipment

Equipment needed to condition the gas may be located at any one of several places in the gathering system; it may be needed at the wellhead, or it may be possible to locate it at a more central point. The equipment described here will be discussed in greater detail in other chapters of this manual.

Separators are vessels that function as a wide place in the pipeline so that the flowing fluids slow down and gravity separates the vapors and solids from the liquids. Some are designed to separate different liquids such as condensate from water. Separators are called by many names, such as scrubbers, traps, knockouts, and drips.

Heaters are usually low-pressure vessels that contain a liquid—most often water—which is heated by a gas burner using fuel from the line. Pipe of sufficient strength is coiled and placed in the hot liquid, and the gas to be heated is passed through the coiled pipe. Gas heaters may be directly or indirectly fired.

Dehydrators prevent hydrate formation by removing water vapor from the gas. The gas is brought into contact with either a liquid or solid desiccant, which takes water vapor from the gas. The desiccant is then regenerated for reuse by applying heat, which drives off the water picked up from the gas.

Compressors are installed to increase the pressure of the gas so that it can flow through the pipeline.

Friction is developed by flow through the pipe; so if the pipeline is very long, additional compressors may have to be installed along the line. Reciprocating piston compressors are the type most often used; the smaller, high-speed units are used for small volumes, the larger, low-speed, integral units for large volumes. Centrifugal compressors are becoming more common, particularly for main gas-transmission lines.

Natural gas that contains corrosive elements must be conditioned so that the flow-system equipment does not deteriorate and become a safety hazard or an economic loss. If small quantities are involved, the conditioning may be by the injection of chemical inhibitors into the gas stream; if the quantities are large, the most economical method may be removal of the contaminant. Removal is usually done with equipment and procedures similar to those used to dehydrate gas.

Gas plant, gasoline plant, and *gas-liquids extraction plant* are all terms applied to facilities that remove some of the heavier or easily liquified hydrocarbon component from the gas so that it can be used separately from the gas. This is only done, of course, where there is justification from an economic standpoint, conservation of natural resources, or the need of the consumer. See Appendix B for a more complete description of the purpose and design of gas-processing plants.

IV

NATURAL GAS AND LIQUID SEPARATION

Petroleum as produced from a reservoir is a complex mixture of hundreds of different compounds of hydrogen and carbon, all with different densities, vapor pressures, and other physical characteristics. A typical well stream is a high-velocity, turbulent, constantly expanding mixture of gases and hydrocarbon liquids, intimately mixed with water vapor, free water, solids, and other contaminants. As it flows from the hot, high-pressure petroleum reservoir, the well stream is undergoing continuous pressure and temperature reduction. Gases evolve from the liquids, water vapor condenses, and some of the well stream changes in character from liquid to bubbles, mist, and free gas. The high-velocity gas is carrying liquid droplets, and the liquid is carrying gas bubbles.

Stated simply, field processing is to remove undesirable components and to separate the well stream into salable gas and petroleum liquids, recovering the maximum amounts of each at the lowest possible overall cost. Field processing of natural gas actually consists of four basic processes:

1. Separation of the gas from free liquids such as crude oil, hydrocarbon condensate, water, and entrained solids
2. Processing the gas to remove condensable and recoverable hydrocarbon vapors
3. Processing the gas to remove condensable water vapor, which under certain conditions might cause hydrate formation
4. Processing the gas to remove other undesirable components, such as hydrogen sulfide or carbon dioxide

CONVENTIONAL SEPARATORS

Separation of well-stream gas from free liquids is by far the most common of all field-processing operations and, at the same time, one of the most critical. A properly designed separator essentially will provide a clean separation of free gases from the free hydrocarbon liquids. A well-stream separator must perform the following:

1. cause a primary-phase separation of the mostly liquid hydrocarbons from those that are mostly gas;
2. refine the primary separation by removing most of the entrained liquid mist from the gas;
3. further refine the separation by removing the entrained gas from the liquid; and
4. discharge the separated gas and liquid from the vessel and insure that no reentrainment of one into the other takes place.

If these functions are to be accomplished, the basic separator design must—

1. control and dissipate the energy of the well stream as it enters the separator;
2. insure that the gas and liquid flow rates are low enough so that gravity segregation and vapor-liquid equilibrium can occur;
3. minimize turbulence in the gas section of the separator and reduce velocity;
4. control the accumulation of froths and foams in the vessel;
5. eliminate reentrainment of the separated gas and liquid;
6. provide an outlet for gases, with suitable controls to maintain preset operating pressure;
7. provide outlets for liquids, with suitable liquid-level controls;
8. if necessary, provide cleanout ports at points where solids may accumulate.
9. provide relief for excessive pressures in case the gas or liquid outlets should be plugged; and

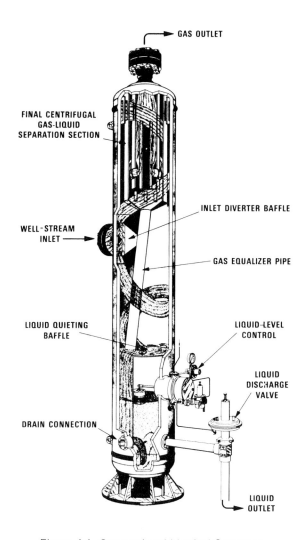

GAS OUTLET

FINAL CENTRIFUGAL
GAS-LIQUID
SEPARATION SECTION

INLET DIVERTER BAFFLE

WELL-STREAM
INLET

GAS EQUALIZER PIPE

LIQUID QUIETING
BAFFLE

LIQUID-LEVEL
CONTROL

LIQUID
DISCHARGE
VALVE

DRAIN CONNECTION

LIQUID
OUTLET

Figure 4.1. Conventional Vertical Separator

10. provide equipment (pressure gauges, thermometers, and liquid-level gauge-glass assemblies) to check visually for proper operation.

The process equipment and conditions downstream of a separator will usually dictate the necessary degree of separation and the actual vessel design. Ideally, gases and liquids should come to full equilibrium in the separator, but a compromise must usually be made between the degree of separation achieved and cost of the installation.

The following factors must be considered in sizing and selecting a seperator.

1. Liquid flow rate (oil and water), barrels per day and minimum and peak instantaneous.
2. Gas flow rate, million standard cubic feet (MMscf) per day.
3. Specific gravities of oil, water, and gas.
4. Required retention time of fluids within the separator; retention time is a function of physical properties of the fluids.
5. Temperature and pressure at which the separator will operate and design pressure of the vessel.
6. Whether the separator is to be two phase, such as liquid and gas, or three phase—that is, oil, water, and gas.
7. Whether or not there are solid impurities, such as sand or paraffin.
8. Whether or not there are foaming tendencies.

Three basic types of separators are widely used for gas-liquid separation: (1) vertical (fig. 4.1), (2) horizontal (fig. 4.2), and (3) horizontal double barrel (fig. 4.3). Each has specific advantages, and selection is usually based on which one will accomplish the desired results at the lowest cost.

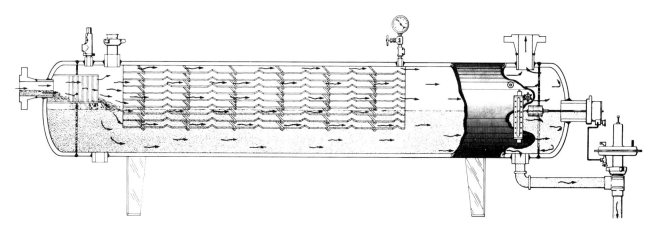

Figure 4.2. Conventional Horizontal Separator

37

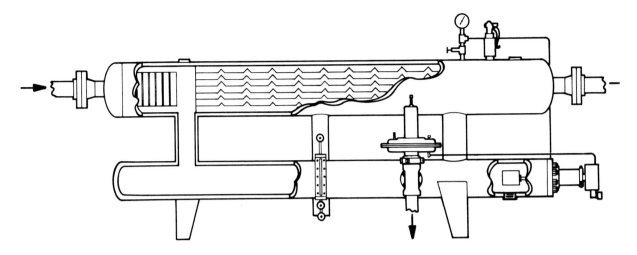

Figure 4.3. Conventional Horizontal Double-Barrel Separator

Vertical

A vertical separator is often used on low to intermediate gas-oil ratio well streams and where relatively large slugs of liquid are expected. It can be fitted with a false cone bottom to handle sand production. A vertical separator occupies less floor space, an important consideration where this might be expensive as on an offshore platform. However, because the natural upward flow of gas in a vertical vessel opposes the falling droplets of liquid, a vertical separator for the same capacity may be larger and more expensive than a horizontal unit. In operation, an inlet diverter spreads the inlet fluids against the vertical separator shell in a thin film and at the same time imparts a centrifugal motion to the fluids. This provides the desired momentum reduction and allows the gas to escape from the thin oil film. The gas rises to the top of the vessel, and the liquids fall to the bottom. Some small liquid particles will be swept upward with the rising gas stream, and these particles are separated by a centrifugal baffle arrangement below the gas-outlet connection.

Horizontal

The horizontal separator may be the best separator for the money. It may be less expensive than the vertical separator for equal capacity. The horizontal separator has a much greater gas-liquid interface area consisting of a large, long, baffled gas-separation section, which permits much higher gas velocities. Horizontal separators are almost always used for high gas-oil ratio well streams, for foaming well streams, or for liquid-from-liquid separators. A horizontal separator is easier to hook up, easier to service, easier to skid-mount. Several separators can be stacked easily into stage-separation assemblies minimizing space requirements. In a horizontal separator, gas flows horizontally and, at the same time, falls toward the liquid surface. Some separators have closely spaced horizontal baffle plates that extend lengthwise down the vessel upon which are evenly spaced baffle plates at a forty-five-degree angle to the horizontal. The gas flows in the baffle surfaces and forms a liquid film that is drained away to the liquid section of the separator. The baffles need only to be longer than the distance of liquid trajectory travel at the design gas velocity.

Some separators use knitted wire-mesh pads, 4 to 8 in. thick, across the gas section of the separator or across the gas-outlet nozzle. Wire mesh can provide good gas-liquid separation but may be plugged by paraffin or solids in the gas stream. Wire-mesh pads should not be used unless the well stream is so clean that no danger of plugging exists.

Double-Barrel Horizontal

A double-barrel horizontal separator has all the advantages of a normal horizontal separator plus a much higher liquid capacity. Incoming free liquid is immediately drained away from the upper section into the lower section. The upper section is filled with baffles, and gas flow is straight through and at higher velocities.

A horizontal three-phase separator (fig. 4.4) is designed to separate oil, water, and gas and has two

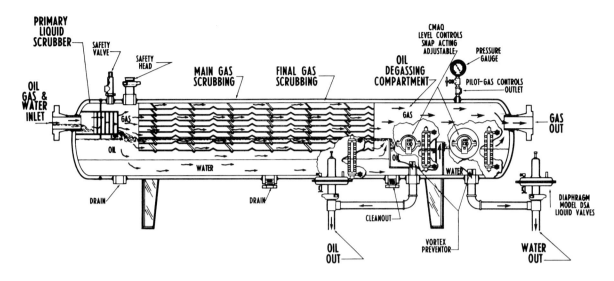

Figure 4.4. Conventional Horizontal Three-Phase Separator

liquid outlets. Three-phase separators are used commonly for well testing and in instances where free water readily separates from the oil or condensate. They are identical to two-phase vessels except for the water compartment and an extra level control and dump valve.

Separator Controls

Liquid level within the separator must be maintained within reasonable limits to prevent discharging liquid out of the gas line or gas out of the liquid line and to assure proper functioning and flow through the separator internals.

Pressure within the separator is usually maintained within a specific pressure range by a gas back-pressure regulating valve.

Temperature within most separators is usually not controlled although there are exceptions, such as low-temperature separation systems.

Safety and protection against overpressure is provided by a pressure-relief valve set at the design pressure of the separator.

FILTER SEPARATORS

Filter separators are designed to remove small liquid and/or solid particles from gas streams. These units were designed specifically to handle those applications where, due to the extremely small particle size, conventional separation techniques employing gravitational or centrifugal force are ineffective. Contaminants of this small size can be removed most effectively by passing the gas through a fine high-quality filtering medium. Several configurations of filter separators are used, depending upon the required efficiency and on whether liquids or solids or both are to be removed. Some filter elements have collection efficiencies of 98 percent of the 1-micron particles and 100 percent of the 5-micron particles when operated at rated capacity and recommended filter-change intervals.

Applications

Figure 4.5 illustrates a simple dry-gas filter separator. Where removal of dry solids is the only requirement, this configuration is used. It has no provisions for collecting or dumping liquids. A dry-gas filter separator is used in lines carrying gas undersaturated with condensable hydrocarbon and water—for example, from a stripping plant. It may be used where liquid occasionally is present if the passage of liquid through the filter is not objectionable. Filter elements are easily changed from the hinged quick-opening enclosure.

When both liquid and solid contaminants must be removed from the gas, a two-compartment vessel shown in figure 4.6 is normally used. Liquid is coalesced, and solid particles are filtered from the gas by a glass-fiber element. Solids are retained by the element, while liquids are agglomerated by wetting the surface of the fiber glass. Liquids are blown off

39

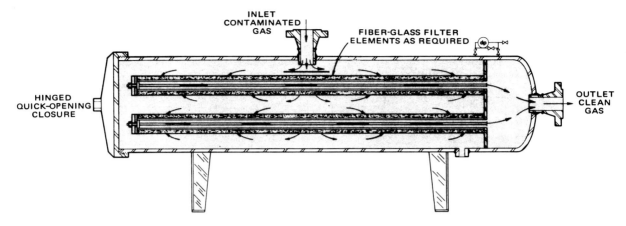

Figure 4.5. Dry-Gas Filter Separator

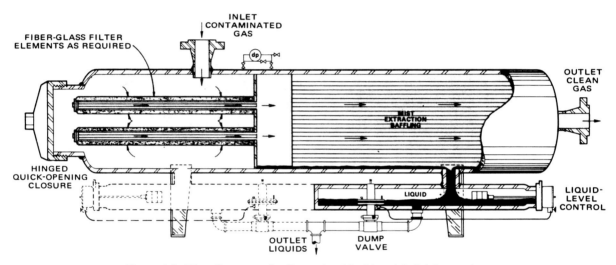

Figure 4.6. Filter Separator for Removing Liquid and Solid Contaminants

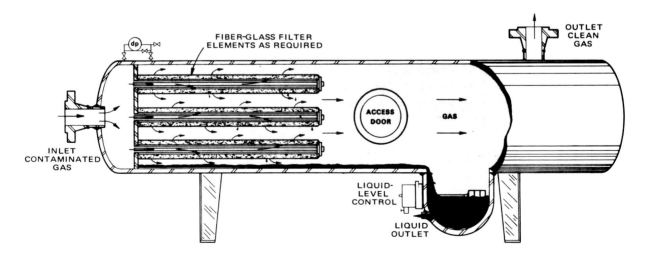

Figure 4.7. Filter Separator for Fog Coalescing and Gas Filtering

40

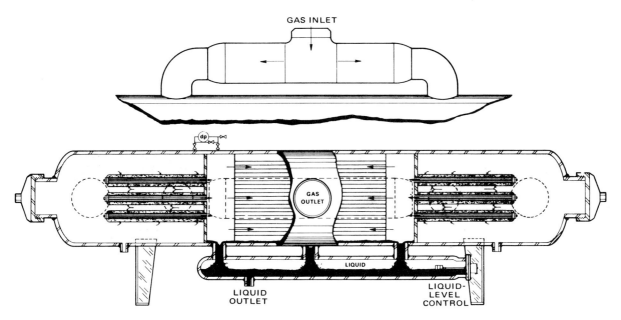

Figure 4.8. Large Capacity Filter Separator

the outlet end of the fiber-glass element in large droplets, which are then easily captured by mist-extracting baffles. Liquids are collected in the lower barrel and dumped by a level-control and valving system. A primary liquid collection lower barrel (dotted in on fig. 4.6) is required where liquid volumes are large or when liquid enters the vessel in slugs.

Figure 4.7 illustrates a filter separator designed for fog coalescing and gas filtering. If the major problem is fog coalescing, the gas flow through the elements may be from inside out, the reverse of normal flow. In this application, the liquid collects in large droplets on the outside of the element and can fall to the bottom of the vessel without being broken up by the high gas velocity. Therefore, no mist-extraction baffling is required. Liquid flows to the catch pot in the bottom of the vessel where discharge is controlled by a liquid-level control.

Figure 4.8 illustrates a large capacity filter separator. Large capacities can often be best handled by a dual vessel arrangement. Inlets are provided at each end with flow toward a common center outlet. Independent filter element sections are provided at each end with quick-opening access closures for filter removal. Liquids are collected in the lower barrel and dumped by a liquid-level control and valve system.

Filter Elements

A standard filter element is made of a perforated metal cylinder with gasketed ends for compression sealing. A fiber-glass cylinder with a ½-in.-thick wall is located outside the metal cylinder. The gas flow is normally from the outside of the cylinder to the center. There is a layer of fabric outside and inside the glass cylinder to prevent the migration of the glass fibers into the gas stream. (See fig. 4.9.) Some

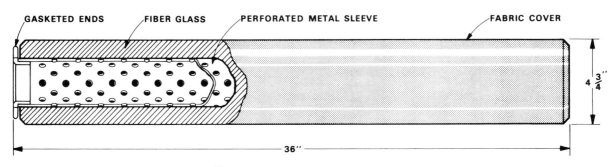

Figure 4.9. Standard Filter Element

TWO-STAGE SEPARATION

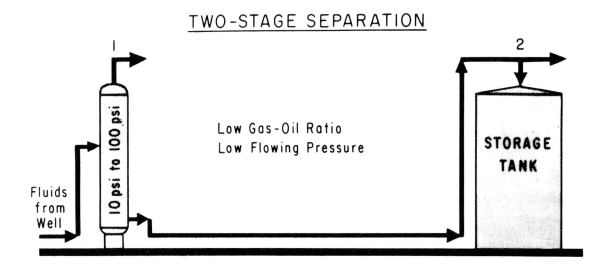

Low Gas-Oil Ratio
Low Flowing Pressure

THREE-STAGE SEPARATION

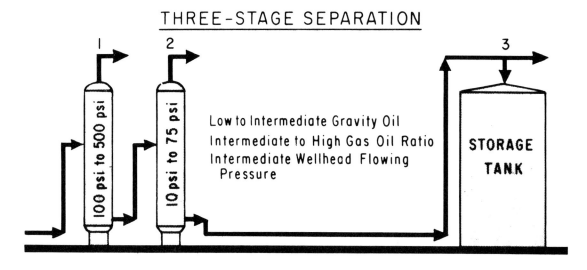

Low to Intermediate Gravity Oil
Intermediate to High Gas Oil Ratio
Intermediate Wellhead Flowing
    Pressure

FOUR-STAGE SEPARATION

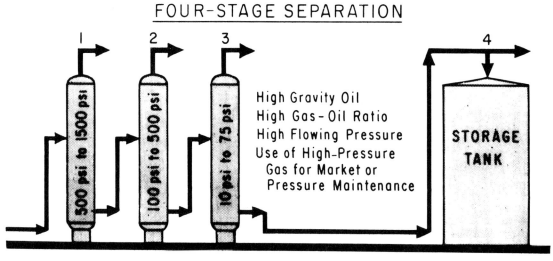

High Gravity Oil
High Gas-Oil Ratio
High Flowing Pressure
Use of High-Pressure
    Gas for Market or
    Pressure Maintenance

Figure 4.10. Stage Separation Flow Diagrams

elements are used with flow from the center to the outside when the primary duty is coalescing of liquid particles. Hoop strength is added to these elements by glass fibers wrapped around the element. Filter elements are securely held over openings in the vessel tube sheet by a center rod. This rod centers each element over its tube-sheet opening and provides the compression for sealing the element between the tube sheet and the plate, which closes the opposite end. The center rod is removable from the vessel for unobstructed access.

Pressure Drop

Filter cartridges offer little pressure drop to the flow of gas when they are clean. Pressure drop through the entire vessel with clean elements will be 2 psi at its rated capacity unless stated otherwise. Approximately half of this pressure drop will be caused by the elements supporting hardware. Filter elements should be changed when the differential pressure across them rises approximately 6 psi above the initial (clean) pressure drop. Never permit the differential pressure to reach the 30 psi element collapse pressure. This pressure can be read directly on the differential pressure gauge, which is provided. A 30 psi pressure gauge encased in a 2,500 working-pressure glass-front case is used for this service.

STAGE SEPARATION

Stage separation is a process in which gaseous and liquid hydrocarbons are separated into vapor and liquid phases by two or more equilibrium flashes at consecutively lower pressures. As shown in figure 4.10, two-stage separation involves one separator and

a storage tank. Three-stage separation requires two separators and a storage tank. Four-stage separation would require three separators and a storage tank. The tank is always counted as the final stage of vapor-liquid separation because the final equilibrium flash occurs in the tank.

The purpose of stage separation is to reduce the pressure on the reservoir liquids a little at a time, in steps, or stages, so that a more stable stock-tank liquid will result. Petroleum liquids at high pressures usually contain large quantities of liquefied propanes, butanes, and pentanes, which will vaporize or flash as the pressure is reduced. This flashing can cause a substantial reduction in stock-tank liquid recovery, depending upon well-stream composition, pressure, temperature, and other factors. For example, if a volatile condensate at 1,500 psig was discharged directly into an atmospheric storage tank, most of it would immediately vaporize leaving very little liquid in the tank.

The ideal method of separation, to attain maximum liquid recovery, would be that of differential liberation of gas by means of a steady decrease in pressure from that existing in the reservoir to that existing in the storage tanks. With each tiny decrease in pressure, the gas evolved would immediately be removed from the liquid. However, to carry out this differential process would require an infinite number of separation stages, obviously an impractical solution. We can make a close approach to differential liberation by using three or more series-connected stages of separation, in each of which flash vaporization takes place. In this manner, the maximum economical amount of liquids can be recovered. Figure 4.11 shows a typical skid-mounted horizontal-stage separation unit.

Figure 4.11. Skid-mounted Two-Stage Separator Unit

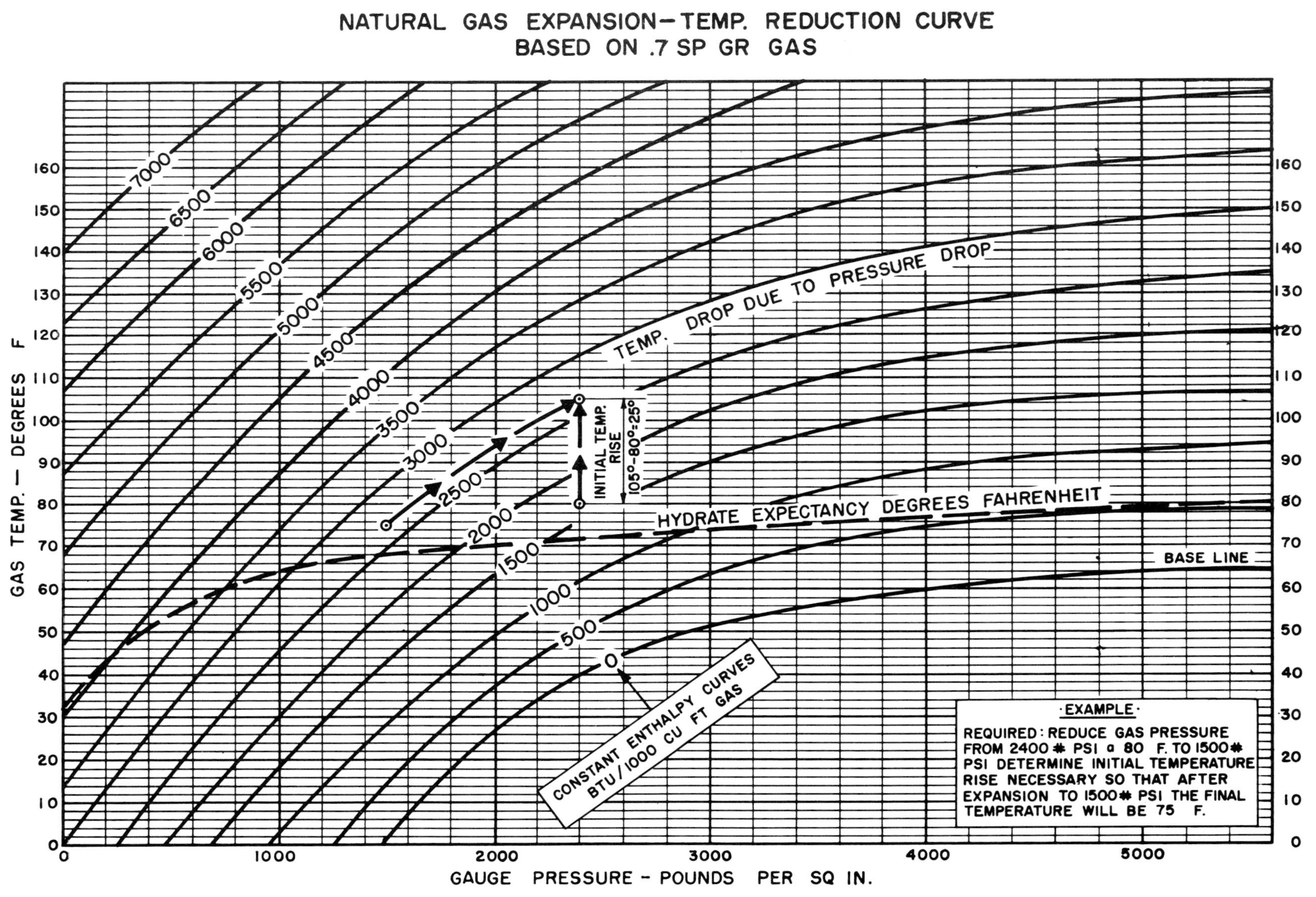

NATURAL GAS EXPANSION–TEMP. REDUCTION CURVE
BASED ON .7 SP GR GAS

Figure 4.12. Natural Gas Expansion–Temperature Reduction Curves

LOW-TEMPERATURE SEPARATION

Low-temperature separation, probably the most efficient means yet devised for handling high-pressure gas and condensate at the wellhead, performs the following functions:

1. separation of water and hydrocarbon liquids from the inlet well stream;
2. recovery of more liquids from the gas than can be recovered with normal-temperature separators; and
3. dehydration of gas, usually to pipeline specifications.

The first low-temperature units were developed and placed in operation in 1948, to dehydrate gas at the wellhead in remote locations so that high-pressure gas and condensate could be gathered at central locations without the problems of hydrate plugging.

Essentially, the low-temperature separation process is one of intentional hydrate formation and controlled melting. The inlet gas is cooled by expansion, due to pressure reduction, causing water and liquid hydrocarbons to condense; if hydrates are formed, they are quickly melted. Dry gas, condensate, and free water are then discharged from the vessel under controlled conditions. Since this separation system permits operating conditions well below hydrate-formation temperatures, the recovery of hydrocarbon liquids is much higher than that possible for conventional separation. Condensation of a higher percentage of the water vapor also is accomplished, resulting in dehydration of the gas.

Since low-temperature separation is achieved by the cooling effect of gas expansion—that is, pressure reduction across a choke—how much pressure drop is needed to obtain adequate cooling? Generally speaking, satisfactory operation and dehydration to pipeline specifications can be accomplished with a pressure differential as low as 1,000 psi, provided the temperature upstream of the choke can be controlled to near the hydrate point.

The actual minimum differential between flowing and line pressures depends primarily upon the pressure range in which the hydrate temperature falls. Noting the pressure-temperature drop chart (fig. 4.12), it can be seen that a 1,500 psi drop from 3,000

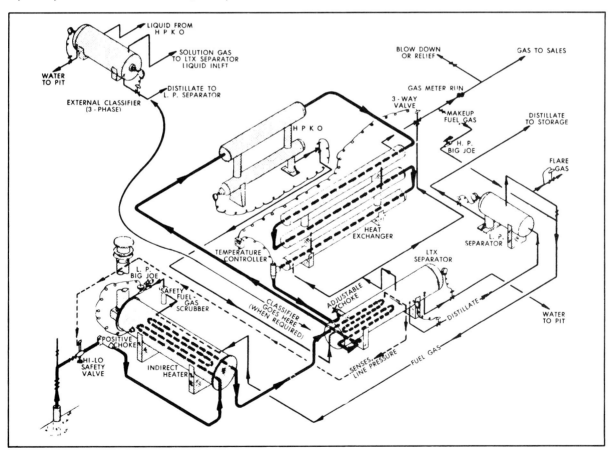

Figure 4.13. Schematic Flow Diagram of a Low-Temperature Separation Unit

psi at 120 F will provide a final temperature of only 78 F, while a 1,500 psi drop from 2,000 psi at 120 F will give a final temperature of 49 F. The actual temperature drop per psi of pressure drop will not necessarily correspond to that shown on the chart; composition of the gas stream, flow rate, liquid rates, bath temperature, and ambient temperature will affect the actual temperature drop. Water dew points will average 10 F to 15 F below the indicated temperature due to the adsorptive effects of the hydrates on the vapor-phase water.

Low-Temperature Systems

Low-temperature systems may be discussed in two categories: (1) those that use hydrate inhibitors and (2) those that do not. The same principle of refrigeration by expansion is utilized in both systems, but the details differ sufficiently to warrent separate discussion. A third category of low-temperature separation systems, called the mechanical refrigeration system, operates just the same as the first two; the only difference is that the expansion of high-pressure gas is not used to cool the inlet well stream. Instead, a separate closed-cycle propane or ammonia refrigeration system is used to cool the well stream before separation. A mechanical refrigeration system is often used when well-stream flowing pressures are too low to obtain adequate pressure differential and expansion for cooling.

A schematic flow diagram of a typical low-temperature separation system, using no hydrate inhibitor, is shown in figure 4.13. The basic equipment is as follows:
1. inlet well-stream heater (optional);
2. inlet free-liquid separator;
3. heat exchanger, which changes cold sales gas to warm inlet gas;
4. choke or other expansion valve on inlet to low-temperature separator;
5. low-temperature separator with heat-exchange coil in liquid section;
6. flash separator; and
7. piping and controls.

Depending upon the flowing temperature of the inlet well stream, the inlet well-stream heater may or may not be required. The well stream must be warm enough to melt any hydrates that may form in the bottom of the low-temperature separator. Following the flow diagram, assuming a heater is required, the inlet well stream flows through the heater coil and then through a coil located in the bottom of the low-temperature separator (fig. 4.14).

The well stream provides heat to warm the separator liquids and melt hydrates that may have formed. In passing through this coil, the well stream is partially cooled, causing condensation of some water and liquid hydrocarbons.

The well stream then flows to a high-pressure liquid knockout, where the liquids may be handled in one of three ways.
1. Water only removed in the knockout; the condensate goes across the choke with the gas.
2. All liquids removed in the knockout and then dumped into the liquid section of the low-temperature separator.
3. All liquids removed in the knockout, with the water being dumped to disposal and the condensate dumped to the liquid section of the low-temperature separator.

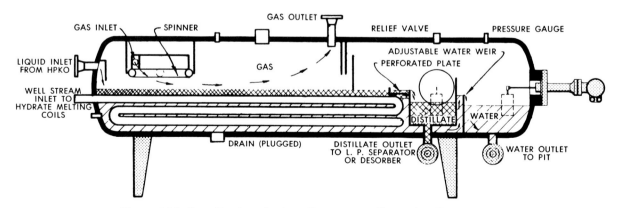

Figure 4.14. Cross Section of a Low-Temperature Extraction Separator

Figure 4.15. Skid-mounted Low-Temperature Separation System

The well stream then leaves the knockout and flows through the tube side of a sales-gas/inlet-gas heater exchanger, where the inlet well stream is cooled by the cold gas leaving the low-temperature separator. Temperature control of the well stream is very critical at this point since too low a temperature will result in hydrate formation and plugging of the exchanger tubes or the choke. Too high a temperature will result in inadequate dehydration of the gas and reduced recovery of condensate.

After being properly cooled, the well stream flows through the choke expanding directly into the low-temperature separator, dropping from inlet pressure to sales-line pressure, and cooling as a result of the expansion. This cooling causes some of the hydrocarbon gases and most of water vapor to liquefy and, in most cases, causes hydrates to form. The hydrates, water, and condensate are deposited in the warm liquid bath where the hydrates are melted and the condensate and water are separated.

The cold, dehydrated gas leaves the low-temperature separator, flowing through or partially bypassing the previously mentioned gas-to-gas heat exchanger. The volume of cold sales gas flowing through the shell side of this exchanger is determined by a three-way proportioning valve. The proportioning valve is controlled by a temperature controller whose bulb is in the well stream immediately upstream of the choke. The excess cold gas, which is not required for cooling the well stream, is diverted through the bypass piping and then recombined with the cooling gas. The gas then flows through an orifice meter to the sales line.

The hydrocarbon liquids, which were dumped from the inlet free-liquid knockout, are dumped into the liquid section of the low-temperature separator. The evolved solution gas from these hydrocarbon liquids then joins the main gas stream, which has already expanded across the choke.

The relatively long residence time of the hydrocarbon liquids in the low-temperature separator, plus the heat from the warm hydrate-melting coils causes the condensate to reach a considerable degree of stabilization in the low-temperature vessel. The hydrocarbon components, which may have been vaporized by the heating coil, will be recondensed by the cool gas and will remain in the liquid phase.

Water from the low-temperature separator is discharged directly to disposal facilities. All controls and instruments are operated from a pilot gas system using dehydrated gas, which has been warmed in the heater.

Hydrocarbon liquids from the low-temperature separator are usually discharged to a second-stage low-pressure flash separator to conserve liquid volume. Alternately, some installations may use a condensate stabilizer if a controlled vapor-pressure product is required. Condensate would then be discharged from the flash separator and stabilizer to the liquid storage tanks.

In order to conserve gas, fuel for the inlet heater and instrument-supply gas is withdrawn from the flash gas stream leaving the low-pressure separator. Well streams producing 10 bbl per MMscf or more will usually provide sufficient flare gas to operate the heater and controls. A photograph of a skid-mounted low-temperature separation system is shown in figure 4.15.

The gas volume flowing through a low-temperature unit may be regulated by a manually controlled or pilot-operated choke. Manually controlled chokes are normally used where both volume of flow from the well and pipeline delivery pressures are constant. If delivery-line pressure does not change and wellhead pressures remain constant, the rate of flow through the unit will be constant for a given choke setting. Where wellhead pressures are variable or when pipeline pressures vary from day to day, it is often necessary to regulate flow through the unit by a pilot-operated choke or pressure reducing regulator.

Glycol-Injection System

The system in which a hydrate inhibitor is used provides for injection of glycol prior to the expansion of the gas through the choke. This eliminates the necessity for a heating coil in the bottom of the low-temperature separator. It is necessary in this system to inject sufficient glycol to prevent hydrate formation ahead of the choke and in the low-temperature separator. In this system, the glycol is separated and sent to a regenerator where the absorbed water is driven off by heating so that the glycol can be reused.

The volume of glycol to be injected into the well stream depends upon the inlet conditions of pressure and temperature; these factors determine the amount of water present. In systems with inlet temperatures in the range of 90 F and pressures of 1,700 psi, approximately 5 gal of 85 percent (by weight) diethylene glycol is required per 1 million cu ft of gas. This amount increases with increasing temperature and decreases with increasing pressure. Most installations are operated with injection of 6 to 10 gal per 1 million cu ft.

Glycol losses in the system occur in three ways: (1) solution in condensate, (2) carry-over in the gas stream, and (3) losses in the regenerator. The solubility of diethylene glycol in condensate is approximately 1 gal per 100 bbl at 60 F and 100 lb. Carry-over in the low-temperature separator and in the regenerator is dependent upon the design of the equipment to a large extent but is normally less than 0.2 gal per 1 million cu ft.

In some systems, glycol is injected into the gathering system at or near the wellhead, and the gathering system is operated with minimum hydrate troubles to a central point in the field for separation of condensate and gas. In this type of system, the volume of glycol required may be reduced by the installation of a free-liquid knockout prior to the glycol-injection point to remove free water produced from the well.

When there is no provision for melting hydrates in the low-temperature separator of a glycol-injection system, some operators use a hydrate detector between the choke and the low-temperature separator to prevent plugging the separator. A hydrate detector is simply an enlargement in the pipe filled with baffles or ceramic rings. In the event all free water is not absorbed by the injected glycol, hydrates will form and accumulate in the hydrate detector causing a pressure drop across it. This would indicate the need to reduce the flow of gas or to increase the amount of glycol injected until all free water is removed and the hydrates in the detector are melted.

CONDENSATE STABILIZATION

One of the problems in using low-temperature separation units of both the mechanical and glycol-injection types is high stock-tank vapor loss. These losses are the result of vaporizing appreciable quantities of liquid propane and butane with dissolved methane and ethane, which are liberated when the pressure on the liquid is reduced from the low-temperature separator to storage pressure. When these light ends vaporize in a stock tank, they carry some of the heavier hydrocarbons with them to be burned or lost in the atmosphere. *Stabilization* is a means of removing these lighter hydrocarbons from the liquid present in the bottom of the low-temperature separator with a minimum loss of heavier hydrocarbons. Stabilization results in a larger volume of stock-tank liquids available for sale.

A schematic diagram of glycol-injection system with a stabilizer is shown in figure 4.16. The stabilization system consists of a vertical vessel, which may be packed with ceramic rings or fitted with trays spaced from 12 to 24 in. apart inside the vessel. The liquid in the lower section of the tower is heated by an indirect heater or steam coils. The cold condensate from the bottom of the low-temperature separator flows directly to the top of the stabilizer or may flow to a flash tank prior to entering the stabilizer. From the flash tank, the gas is routed to become fuel or to be disposed of by other means, and the liquid is taken to the top of the stabilizer. Some of the light hydrocarbons are vaporized and pass from the top of this vessel to the gas-sales, fuel, or vent line. The liquid hydrocarbons flow through the packing or down the trays, absorbing some of the heavier

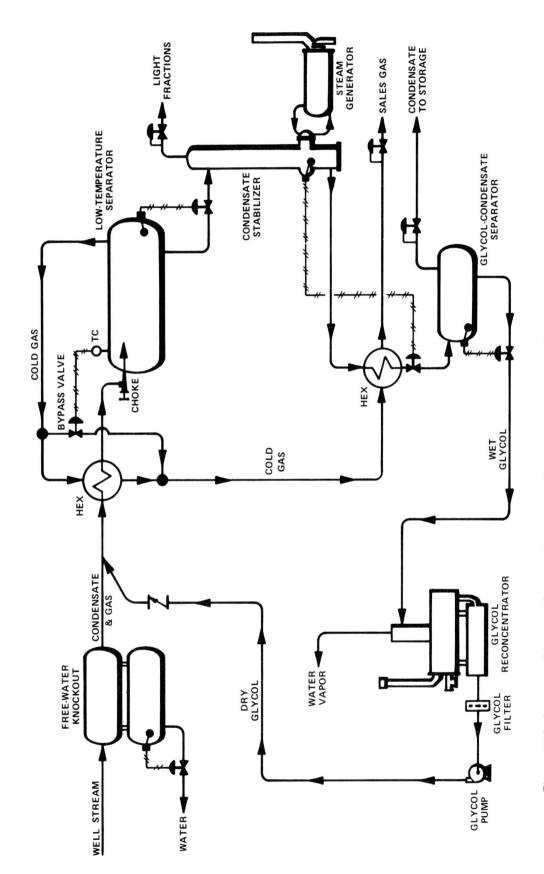

Figure 4.16. Schematic Diagram of a Low-Temperature Separation System with Glycol Injection and Condensate Stabilization

gaseous hydrocarbons, which have been vaporized at the bottom of the vessel. At the bottom of the vessel, the heat added from the heater or reboiler vaporizes most of the lighter hydrocarbons. After being cooled, the stabilized liquid flows to storage and the lighter ends flow upward to be reabsorbed or to leave the top of the vessel. (See fig. 4.17.) The amount of additional condensate that can be recovered by stabilization is dependent upon the pressure and temperature at which the low-temperature separator is operated and the composition of the gas being processed.

Additional hydrocarbon liquid recovery that can be gained by use of low-temperature separation as compared to conventional separation depends on the exact composition of the well stream and upon the operating temperature of the low-temperature separa-

tor. In the absence of an analysis, the increased recovery may be estimated by assuming that stock-tank liquid recovery will be increased about 0.5 bbl per MMscf for each 10 F decrease in separator temperature. For example, if separation at 80 F produces 10 bbl per MMscf, then separation at 20 F would produce an additional 3 bbl per MMscf. An additional increase of about 10 percent might be expected if the low-temperature liquid were then stabilized. For richer gas streams, say in the order of 50 bbl per MMscf, the increased liquid recovery would be about 0.75 bbl per MMscf for each 10 F decrease in separator temperature. These figures are merely rules of thumb and are only an approximation of the actual increased recovery. Figure 4.18 illustrates the effect of separation temperature on liquid recovery.

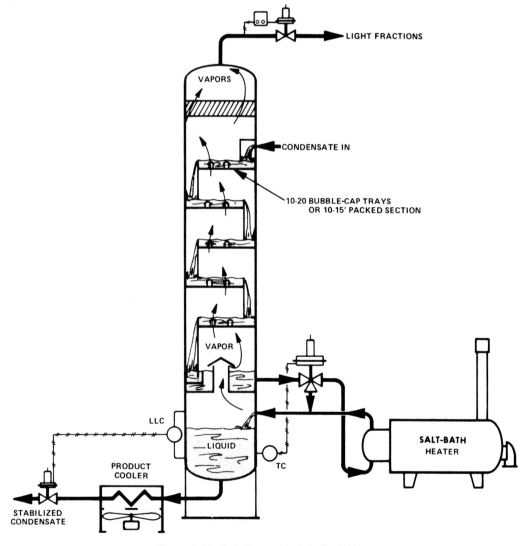

Figure 4.17. Stabilizer with Salt-Bath Heater

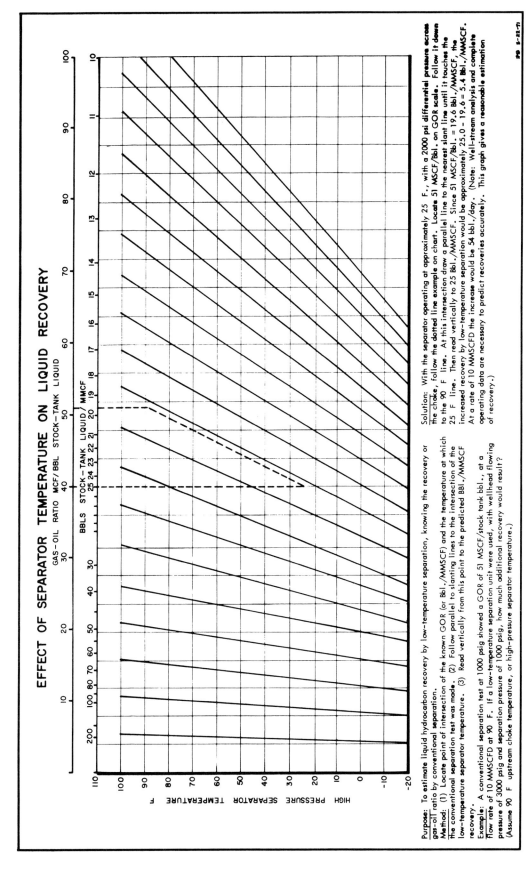

Figure 4.18. Effect of Separator Temperature on Liquid Recovery

51

V

HYDRATES

Most natural gas contains substantial amounts of water vapor at the time it is produced from a well or separated from an associated crude-oil stream. Water vapor must be removed from the gas stream because it will condense into liquid and may cause hydrate formation as the gas is cooled from the high reservoir temperature to the cooler surface temperature. Liquid water almost always accelerates corrosion, and the solid hydrates may pack solidly in gas-gathering systems, resulting in partial or complete blocking of flow lines.

Hydrates are solid compounds that form as crystals and resemble snow in appearance. They are created by a reaction of natural gas with water, and when formed, they are about 10 percent hydrocarbon and 90 percent water. Hydrates have a specific gravity of about 0.98 and will usually float in water and sink in hydrocarbon liquids. Water is always necessary for hydrate formation as well as some turbulence in the flowing gas stream.

FORMATION OF HYDRATES

The temperature at which hydrates will form depends upon the actual composition of the gas and the pressure of the gas stream. Therefore, the chart shown in figure 5.1 cannot be completely accurate for all gases, but it is typical for many gases. The chart shows the water content in pounds of water per MMscf of saturated gas at any pressure or temperature. The dotted line crossing the family of curves shows the temperature at which hydrates will probably form at any given pressure. Note that hydrates form more easily at higher pressures. At 1,500 psig, for example, hydrates may form at 70 F, whereas at 200 psig hydrates will not form unless the gas is cooled to about 39 F. Each curve on the chart shows the water content of a saturated gas at that pressure when the temperature is at any of the various points shown along the bottom of the chart. For example, at 100 psia and 60 F, each 1 million cu ft of gas would contain about 130 lb of water vapor. The same gas at 100 psia and 20 F (instead of 60 F) could only hold 30 lb per MMscf. At 0 F, this gas could only hold 13 lb per MMscf.

It can thus be seen that as a gas is cooled, it can hold less water in the vapor form. Therefore, cooling a gas will cause some of the water vapor to condense with the balance remaining in the gas as water vapor. Pipeline specifications usually require that the water-vapor content of natural gas be 7 lb per MMscf or less in order to minimize the problem of hydrate formation in the transmission lines from the field to ultimate user. In some fields, hydrates form in the tubing and the wellhead valves necessitating the application of heat down the hole to keep the well from freezing up. In most fields, fortunately, the temperature of the gas at the wellhead is 100 F or more; therefore, the hydrate problem does not usually begin until the gas passes through the Christmas tree.

GROUND TEMPERATURES

An all important factor in the movement of gas saturated with water vapor is the retention of the heat, which is in the gas when it is produced. The temperature is lowered at the wellhead when the gas is expanded through a choke to reduce the pressure and control the rate of flow. After passing the choke, the gas enters the gathering lines, which are cooled by

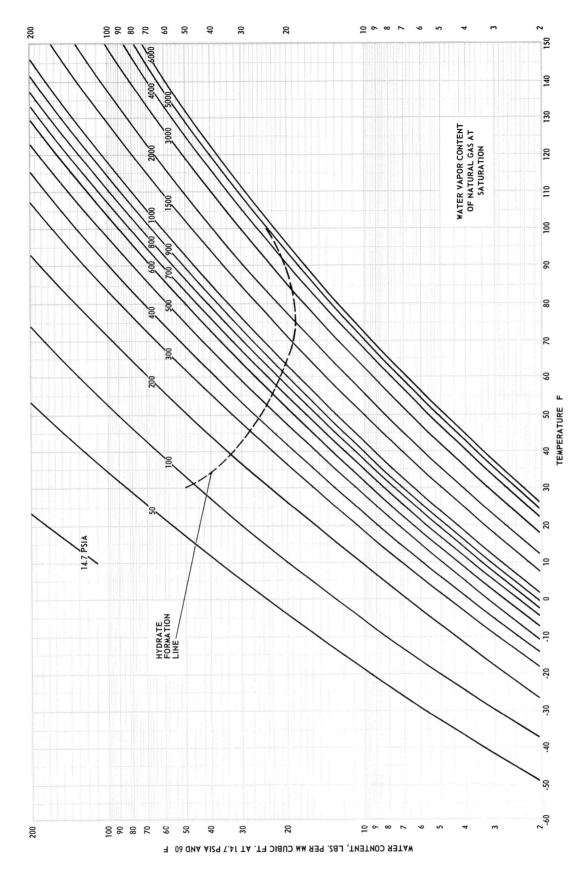

Figure 5.1. Water-Vapor Content of Natural Gas at Saturation

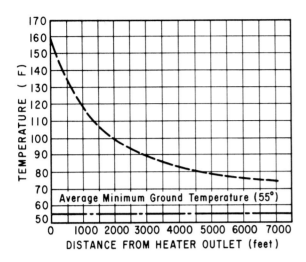

Figure 5.2. Chart of Temperature Drop in Natural Gas Flow Lines

the ground. The effects of the ground temperature are shown in figure 5.2. This curve was prepared from data taken in the Carthage gas field of East Texas. Gas leaving a heater at 160 F had dropped to 75 F by the time the gas had traveled 1½ mi, due to a ground temperature of 55 F. At a higher rate of flow, the gas would travel farther before the formation of hydrates would become a problem. With a line pressure of 900 psi, it could be predicted that hydrates would form at a temperature of about 60 F. Some of the flow lines in this field are as long as 15 mi between well and gasoline plant; so it would appear that this gas will need to be heated again before it reaches the plant.

Minimum ground temperatures at a depth of 18 in. in the various gas-producing regions of the Southwest are as follows:

1. Upper Gulf Coast of Texas and Louisiana, 50 F to 56 F
2. Permian Basin, East Texas, and North Louisiana fields, 45 F
3. Hugoton-Panhandle gas areas, 25 F to 30 F

The figures are approximate and represent minimum values. The ground temperatures may be expected to reach higher levels than this during most of the year, especially in the southern part of Texas and Louisiana. On the other hand, it is probable that in the gas-producing areas of the Rocky Mountain states and in western Canada the frost line is below the 18-in. level. Gas lines in those areas will be buried at correspondingly greater depths.

HYDRATE INHIBITORS

Ammonia, brines, glycol, and methanol have all been used to lower the freezing point of water vapor and thus prevent hydrate formation in flow lines. Low-volume injectors or pumps are used to feed the inhibitor fluid into the gas flow system. Methanol and glycol are the inhibitors most widely used; however, they are expensive. Usually methanol and glycol are used when hydrate problems arise so rarely that the installation of a heater or dehydration equipment is not economically feasible. For example gas-producing operations may be carried on without hydrate trouble except for two or three days in each year. Injection of inhibitors might conceivably be preferred in such cases over investment in additional equipment.

FLOW-LINE HEATERS

In a great majority of the situations where dehydration of the gas is not economical and where some measures must be adopted to control hydrate formation, application of heat is the method used. The reasons for this are obvious. Initial investment is not excessive. Fuel is conveniently available. The heaters operate with a minimum of attention. Where it is desirable to handle the gas by means of temporarily holding the temperature above the ground temperature, heaters seem to offer the best solution. For long-distance transmission, the gas will eventually reach the ground temperature and, for such gas movement, water will ultimately have to be removed.

Indirect Heaters

Most natural gas is produced at relatively high flow-line pressures, ranging from 1,500 to 10,000 psig, and then transferred to pipelines operating at 1,200 psig and lower. Heat must almost always be used to compensate for the natural refrigeration of the gas, which is caused by this pressure reduction. An indirect heater is the most widely used type of heater for natural-gas well streams because it is simple, economical, and, if properly sized, a trouble-free piece of equipment. The heater consists of three basic parts: (1) the heater shell, which is a thin-walled horizontal vessel having removable flanged covers at both ends; (2) a removable fire tube and burner assembly mounted on the lower portion of one of the end covers, and (3) a removable coil assembly mounted on the upper portion of the opposite end cover. The shell and fire tube are designed to withstand only atmospheric pressure; whereas the coil assembly is usually designed to withstand shut-in wellhead pressure.

The arrangement of the equipment of the indirect heater is illustrated in figure 5.3. The high-pressure fluid is introduced to the heater through a choke located on the inlet of the heater, and an expansion takes place immediately inside the heater in the water bath. As the fluid passes through the steel coils, it is heated to a suitable outlet temperature. In operation, the heater shell is filled with a liquid, usually fresh water, completely covering the fire tube and the coil assembly. The water bath is heated by the fire tube, and the coil assembly is heated by the water. Because the water serves as a medium of heat transfer, this type of heater is called in indirect heater. The capacity of such a heater is related to the Btu rating of the burner assembly and fire tube and the effective area of the well-stream coil.

As illustrated, the coil of an indirect heater is located directly above the fire tube. Both the coil and fire tube are immersed in the water bath. Other liquids may be used, but ordinarily the liquid is water. Fundamentally, the fire tube is heat source, and the coils represent points of heat absorption. Hot water has a lower density than cold water and, in a closed vessel, will rise displacing the cold water as the latter settles to the bottom of the vessel due to gravity. Thus, the coils are located above the fire tube. Heat travels from the fire tube to the water immediately surrounding it and causes thermal currents in the water. The direction of these currents can be controlled by the installation of a thermosiphon

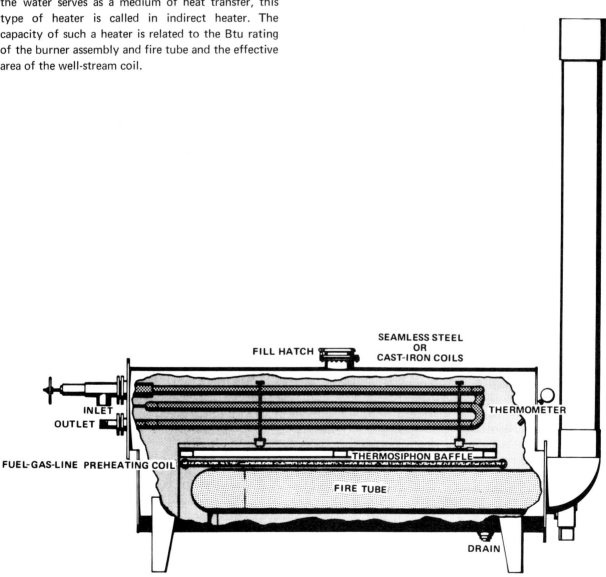

Figure 5.3. Cutaway View of an Indirect Heater

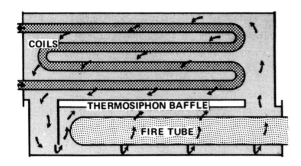

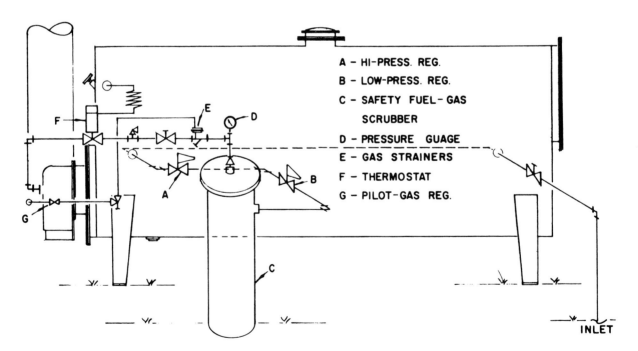

Figure 5.4. Thermosiphon Baffle Arrangement for an Indirect Heater

Coil design is important and especially so when it is considered that the coils of an indirect heater are often manifolded together into bundles with outside piping tying the two bundles together according to requirements.

Most indirect heaters use natural gas as fuel. A more or less standard fuel-gas system is shown in figure 5.5. The inlet gas, usually taken from the first-stage separator, is heated in a small coil immersed in the water bath. Then the gas is regulated to 25 to 40 psig and passed through a safety fuel-gas scrubber, a strainer, the thermostatically controlled valve, the final low-pressure regulator, and the burner assembly. Fuel for the pilot light is taken upstream of the thermostatically controlled valve. An important safety factor is that the main gas valve is prevented from opening if the pilot flame fails for any reason.

baffle, which is illustrated in the cross-section drawing of the indirect heater in figure 5.4. The advantages of directionally controlled thermal currents inside an indirect heater are twofold.

1. The positive velocity of water bath past the fire-tube surface carries away the fire-tube heat at a faster rate and reduces the probability of steam generation thereby reducing and generally eliminating scaling of the fire tube.

2. The overall heat-transfer efficiency of the coil, which is located in the upper part of the heater, is increased because of the increased velocity over the surface of the coils.

When indirect heaters must be located in potentially hazardous areas, the burner assembly and sometimes the stack assembly will be equipped with flame arrestors. These devices permit the heater to draw in sufficient air for proper combustion but prevent explosions or flames within the fire tube from igniting the surrounding potentially hazardous atmosphere.

The following are essential features of a flame arrestor.

A – HI-PRESS. REG.

B – LOW-PRESS. REG.

C – SAFETY FUEL-GAS SCRUBBER

D – PRESSURE GUAGE

E – GAS STRAINERS

F – THERMOSTAT

G – PILOT-GAS REG.

Figure 5.5. Schematic Hookup of Fuel-Gas System for Indirect Heater

1. A bank or honeycomb of small linear passages acts as a mass heat exchanger. These passages are staggered to provide protection and safety.
2. The case holding the arresting element must also withstand a fire-tube explosion and prevent the element from being blown out of the case.
3. The element of the arrestor must be sized to provide sufficient open area for air entrance so that the heater can develop its full Btu rating without requiring forced-draft equipment.

Research has indicated that most heater coil failures result from a combination of corrosion and erosion. Corrosion attacks the surface of the metal, and erosion wipes away the products of corrosion leaving the metal face clean, and therefore, more receptive to new corrosion. It is a vicious and continuous cycle. It has been determined that the rate of corrosion and erosion in the return bends is greater in every case than in the straight-pipe coils attached to the return bends.

As a direct result of field experience and laboratory analysis, manufacturers searched for a way to build into the indirect heater coils a warning signal, if possible. It was found that some refineries made a practice of safety drilling tees, ells, return bends, pipe, and valves. Safety drilling was a simple process of drilling a small hole on the outside of the return bend approximately one-half through the thickness of the metal at a point where erosion and corrosion was likely to be concentrated. (See fig. 5.6.) Such safety drilling provides a place that will issue a warning leak of gas or liquid when approximately half of the metal thickness corrodes or erodes away.

The name *long-nose choke* comes from the fact that the choke body is extended so that the choke orifice may be located within the indirect heater bath. (See fig 5.7.) Thus, a water-saturated gas could be expanded to a temperature below the hydrate-forming temperature of the gas, yet no freezing or plugging of the coil will occur. The use of the long-nose choke results in a greater mean-temperature difference between the coil fluid and the water bath. Since the coil area required to perform a particular heat exchange job is inversely proportional to the mean-temperature difference, the extra cost for the long-nose choke over the standard choke can, in most cases, be paid for in lower coil area requirements.

Other Bath Solutions

While most indirect heaters are of the water-bath type, other liquids and solutions warrant mention.

1. The calcium chloride water solution with inhibitor is used generally to prevent freezing of the water bath. Fouling of the coils will occur if deposition of calcium chloride occurs.

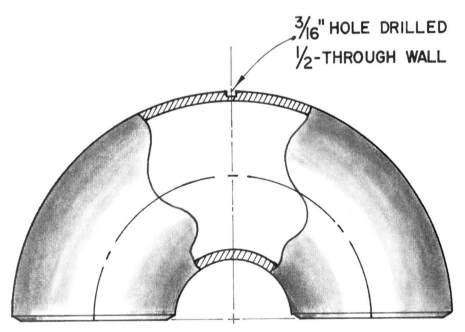

Figure 5.6. Safety Drilled Return Bend for an Indirect Heater

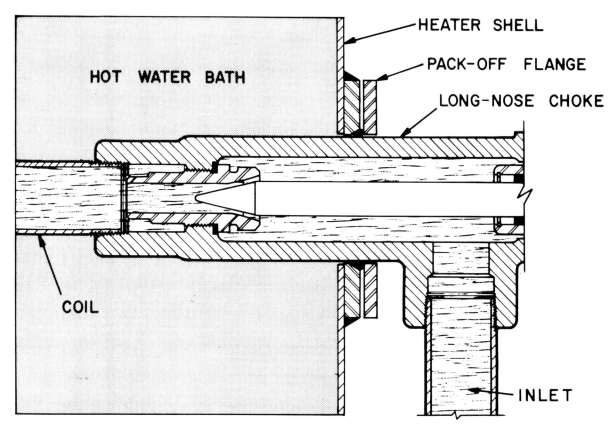

HEATER SHELL

PACK-OFF FLANGE

LONG-NOSE CHOKE

HOT WATER BATH

COIL

INLET

Figure 5.7. Long-Nose Choke Arrangement for an Indirect Heater

2. The glycol formulated antifreeze with inhibitor has glycol in the bath to reduce the heat-transfer rate from the bath to the coil in accordance with the glycol to water ratio and the bath temperature. Should the bath be half glycol and the bath temperature unchanged, the coil outside heat-transfer coefficient may be reduced as much as 20 percent. However, the bath temperature may be increased to 200 F at sea-level pressure without boiling thus compensating somewhat with a higher mean-temperature difference for the loss in heat-transfer coefficient.

3. The steam-bath temperature is 245 F at 15 psig and sea-level atmospheric pressure. Smaller coil area is possible due to—

a. the greater mean-temperature difference;

b. a larger coil outside heat-transfer coefficient.

Insulation for the heater should be considered, and the ASME code stamped units should be employed. In some areas, the ASME vessel is mandatory.

4. The molten salt bath is employed generally in process plants, dehydration units, hydrocarbon recovery plants, and so forth and is not considered applicable for ordinary gas- or oil-producing operations because the high melting temperature—about 600 F—of the salt bath is considerably above maximum temperature requirements for the usual producing operations.

VI

DEHYDRATION OF NATURAL GAS

The term dehydration means removal of water. This section will discuss those processes whereby water vapors—and certain other vapors as well—are removed from the gas by either absorption or adsorption; these processes will be referred to as dehydration. Water vapor may be removed from natural gas by bubbling the gas countercurrently through certain liquids that have a special attraction or affinity for water. When water vapors are removed by this process, the operation is called *absorption.* There are also solids that have an affinity for water. When gas flows through a bed of such granular solids, the water is retained on the surface of the particles of the solid material. This process is called *adsorption.* The vessel in which either absorption or adsorption takes place is called the contactor or sorber. The liquid or the solid having affinity for water and used in the contactor in connection with either of the processes is called the desiccant.

The term *dew point* means the temperature at which natural gas at any specified pressure is saturated with water vapor. Saturated means that the gas contains in vapor form all the water possible at the specified pressure and temperature.

There are two major types of dehydration equipment in use at this time, namely the *liquid-desiccant* dehydrator and the *solid-desiccant* dehydrator. Each has its special advantages and disadvantages and its own field of particular usefulness. Practically all the gas moved through transmission lines is dehydrated by one or the other of these two methods.

DEW-POINT DEPRESSION

Hydrates do not form in a gas line unless the gas is saturated and contains still more water that, since it cannot be absorbed, takes the form of free water. At any specified pressure, hot gas takes more water vapor to reach the saturation point than does cool gas. This means that cool gas that is saturated and has some free water in addition will absorb all the free water when heated sufficiently at the same pressure. If heated above this point, cool gas will not only take up all the free water as water vapor and so prevent hydrate formation but will be undersaturated—that is, will be capable of absorbing more water vapor than there is in the gas. For example, gas at 500 psia and 60 F at the saturation point, contains 30 lb of water per 1 million cu ft. The dew point of this gas is 60 F. Suppose this gas is going to be moved to New York in a transmission line with a temperature of 20 F. The saturation point will then be 7 lb of water per 1 million cu ft. The original 30 lb of water, if left in the gas, will then exist in the form of 7 lb of water vapor and 23 lb of free water per 1 million cu ft, if the pressure remains the same. This free water is a potential source of hydrates to freeze and plug the line. Suppose the gas is processed in a dehydration unit and the dew point is depressed 50 F. This means that no free water will exist in the gas until the temperature is lowered more than 50 F from the original temperature of 60 F or until the temperature goes to 10 F or lower. Gas at 500 psia and 10 F contains about 5 lb of water vapor per 1 million ft. Recalling that this gas originally contained 30 lb of water vapor per MMcf, it will be necessary to remove 25 lb of water from each 1 million cu ft in order to depress the dew point 50 F. In principle, this is the job of the dehydration unit. The problem given above has been stated in oversimplified form purposely to establish the principle of operation. Actual operating problems are not so simple.

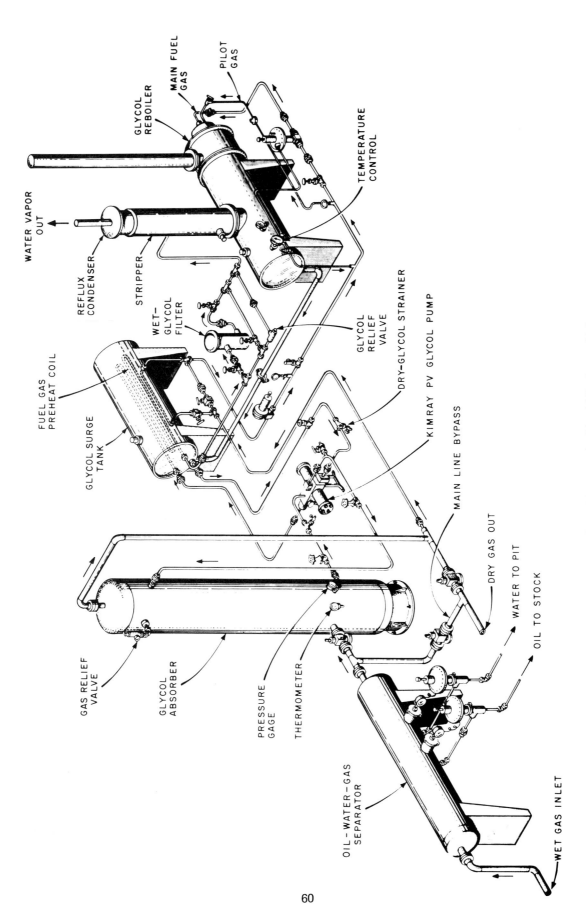

Figure 6.1. Flow Diagram of a Liquid-Desiccant Unit

60

LIQUID-DESICCANT DEHYDRATORS

The desiccant used in the liquid-type dehydrator is usually a solution of one of the glycols, diethylene glycol or triethylene glycol, abbreviated DEG or TEG respectively. The method of operation is the same, TEG is a more recent development in the field than DEG. The following discussion is based on the use of TEG as the desiccant since it is a superior material. It is more easily regenerated to 98 to 99 percent concentration, has a higher (about + 40 F) decomposition temperature, and is subject to lower vaporization losses than DEG.

The following definitions apply to the process description and to the flow diagram in figure 6.1.

Wet gas is gas containing water vapor prior to contacting glycol in the absorber.

Dry gas is gas leaving the absorber after contacting glycol.

Desiccant is a drying or dehydrating medium; here, the desiccant is a triethylene glycol-water (TEG) solution.

Lean Solution is a glycol-water solution whose glycol concentration ranges from 95 to 99 percent by weight. A lean solution can be the solution passing from the reboiler via the pump to the sorber, a reconcentrated solution, or TEG supplied in sealed drums.

Rich solution is a water-rich solution whose glycol content is less than 95 percent by weight or glycol solution that has contacted wet gas in the sorber.

Natural gas at line temperature and pressure enters near the bottom of the absorber (fig. 6.1 and 6.2) and rises through the column where it is intimately contacted by a lean glycol solution flowing downward across bubble trays. Here the gas gives up its water vapor to the glycol. Leaving the top tray, the gas passes through mist-extractor elements, sweeps the glycol-cooling coils located in the upper end of the absorber, and passes to the pipeline. A small quantity of this dry gas is withdrawn from the absorber discharge for use as fuel and instrument gas.

The lean glycol solution enters at the top of the absorber and flows through coils where it is cooled by the dehydrated gas. From the cooling coils, the glycol is discharged in intimate contact with the ascending gas; this action dehydrates the gas and dilutes the glycol at the same time. The dilute solution collects in the base of the absorber from which point it is discharged to the reconcentrator or reboiler. Enroute, the solution is heated in a coil immersed in the glycol surge tank and then discharged into the stripper, which is normally a packed column. The reconcentrated glycol solution accumulates in the reboiler where it reaches maximum temperature, overflows into the glycol surge tanks where it is partially cooled by heat exchange to the dilute glycol in the coil, and flows by gravity to the pump suction. The pump discharges the concentrated solution into the cooling coil in the absorber, thus completing the cycle. Figure 6.3 pictures a small glycol unit.

A dilute glycol solution can be reconcentrated simply by heating it in an open pot to drive off the water as steam. However, a substantial quantity of glycol vapor would be driven off with the water, and these losses are sufficiently large to make the use of a simple boiler economically undesirable. Since water and triethylene glycol have widely varying boiling points (212 F and 549 F), the two substances can be separated easily by fractional distillation. This is accomplished in the packed column or stripper mounted atop the reboiler. Within this column, water-rich vapor from the reboiler rises in intimate contact with descending glycol-rich liquid from the absorber. Between the two phases there is a continuous exchange of heat and materials causing the glycol vapor to condense and liquid water to vaporize as equilibrium is approached. At the top of the column the vapor is virtually pure water, while there is very little water in the glycol at the bottom. By condensing water in the top of the column with an atmospheric heat exchanger or a reflux cooling coil through which the cool, rich glycol circulates before entering the still proper, sufficient liquid, known as *reflux*, is provided for proper fractionation. As long as the temperature at the top of the column ranges from 210 F to 212 F, glycol losses are minimized.

In order to understand the varying conditions under which most liquid-desiccant dehydrators operate, it is necessary to consider the effect of four major operating variables: (1) gas pressure, (2) gas temperature, (3) solution rate, and (4) solution concentration. At constant temperature, the lower the pressure is the higher the water content of the inlet gas will be. At constant pressure, the higher the temperature is the higher the water content of the inlet gas will be. Over a normal pressure range up to 1,200 psig, about 2 gal of glycol must be circulated for every 1 lb of water removed at the 55 F dew-point depression. This quantity is based on equilibrium conditions and a 95 percent glycol solution. Greater dew-point depressions can be attained by increasing

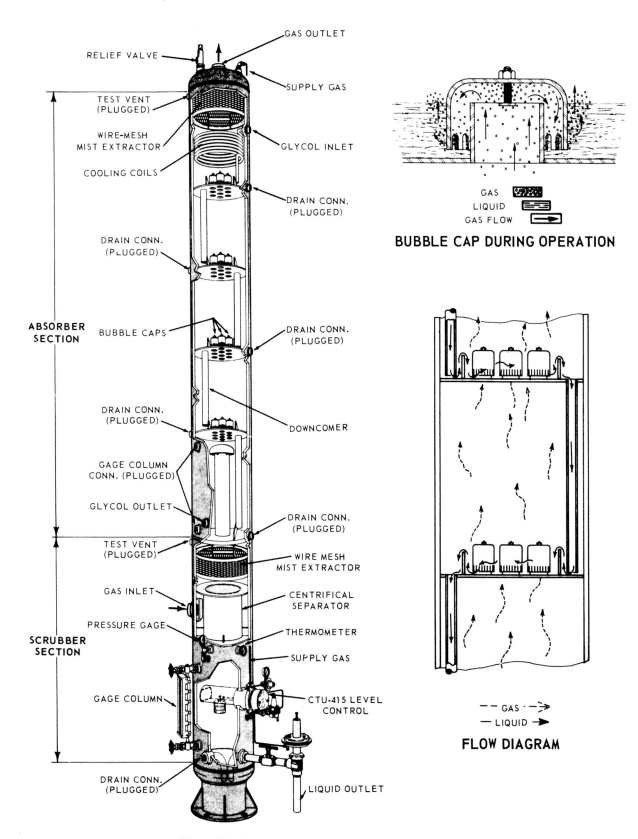

GAS OUTLET

RELIEF VALVE

SUPPLY GAS

TEST VENT
(PLUGGED)

WIRE-MESH
MIST EXTRACTOR

GLYCOL INLET

COOLING COILS

DRAIN CONN.
(PLUGGED)

DRAIN CONN.
(PLUGGED)

ABSORBER
SECTION

BUBBLE CAPS

DRAIN CONN.
(PLUGGED)

DRAIN CONN.
(PLUGGED)

DOWNCOMER

GAGE COLUMN
CONN. (PLUGGED)

GLYCOL OUTLET

DRAIN CONN.
(PLUGGED)

TEST VENT
(PLUGGED)

WIRE MESH
MIST EXTRACTOR

GAS INLET

CENTRIFICAL
SEPARATOR

PRESSURE GAGE

THERMOMETER

SCRUBBER
SECTION

SUPPLY GAS

GAGE COLUMN

CTU-415 LEVEL
CONTROL

DRAIN CONN.
(PLUGGED)

LIQUID OUTLET

GAS
LIQUID
GAS FLOW

BUBBLE CAP DURING OPERATION

-- GAS ·
— LIQUID

FLOW DIAGRAM

Figure 6.2. Glycol Absorber Tower with Scrubber Section

Figure 6.3. Glycol Liquid-Desiccant Dehydrator Unit

63

the circulation rate to the upper limit of the design capacity of the equipment. Also, as the circulation rate exceeds 7 gal per 1 lb of water content, the effect on dew-point depression is greatly diminished. In general, the higher the glycol concentration is the greater the dew-point depression will be. Whereas glycol concentrations of 95 to 96 percent will give dew-point depressions of 55 F, concentrations of 99 percent will give depressions of 65 F, assuming the circulation rate remains constant. A reboiler temperature of about 350 F will give a glycol concentration of 95 to 96 percent, and a reboiler temperature of about 375 F will give a glycol concentration of 97 to 99 percent. Both these values are true at atmospheric pressure. Some manufacturers are marketing high-concentration glycol regeneration units, which permit glycol dehydrators to compete with the dry-desiccant units for low dew points. Dew-point depressions of 130 F to 140 F are now possible with glycol dehydrators using triethylene glycol and the high-concentration glycol regenerators. Triethylene glycol (TEG) concentrations of 99.1 to 99.995 percent TEG can be obtained with the stripping-gas feature.

Most operating problems encountered in the operation of glycol dehydration plants can be related to the contamination of the glycol by foreign materials and the breakdown of the glycol at elevated temperatures. The first place in the system for removal of contaminants is in the inlet wet-gas scrubber. This vessel should be located as closely as possible to the absorber. Its function is to remove water, hydrocarbon condensate, crude oil, lubricating oil, and any pipeline dust or dirt that comes along with the gas. This vessel should be of adequate size and maintained in the proper working order; if light oils and condensate are carried into the absorber, they eventually find their way to the reconcentrator where they are likely to flash. The flashing of these elements may cause fracturing of the packing in the still and blow glycol and hydrocarbons out of the top, resulting in a serious fire hazard. Heavy oils will eventually find their way to the surge tank where they must be drained off. Some oils form an emulsion with the glycol. When this emulsion reaches the regenerator, light fractions of the oil are driven off overhead with the water. In several cases, fires were started from these vapors coming into contact with the fire in the reboiler. This can be prevented by carrying the vapors away from the top of the still column through a downward sloping line into a remotely located separator that can draw off the water separately from the gasoline. Care must be taken to prevent this line from freezing. A glycol filter is usually installed to remove dirt, scale, rust, and reaction materials. This filter is usually of the waste-pack or cartridge type, which can be removed and replaced with the plant in operation. One type of filter sometimes used has an element of cloth, fiber glass, or cotton in the form of a replaceable sock surrounded by a stainless steel wire mesh. It is located upstream of the motor valve controlling the liquid level in the absorber since gas released due to the pressure drop through the motor valve would nullify its operation. The filter in this location must be kept clean. A dirty filter raises the back pressure and the liquid level in the absorber, causing the motor valve to open wider until the pressure drop is taken across the filter instead of the motor valve with a resultant bursting of the filter sack.

Glycols are difficult liquids to contain and to pump. The liquid continually leaks around packing and soon spreads around the entire pump area if not properly handled. There is also a tendency on the part of inexperienced operators to tighten the packing too tightly. The leak problem can be overcome by using a good packing and by providing a drainage trough around the pump leading to a central sump from which the glycol can be gathered and returned to the main reservoir. Metallic packing with a soft babbit center combined with graphite and mica has proved successful. This is a self-lubricating packing that limits the leakage to one or two drops every several minutes.

The reboiler is usually direct fired or steam heated with the glycol level maintained by an overflow pipe to the surge tank or a liquid-level controller. When using level controllers, the float is normally located in the reboiler and the control valve is on the absorber. This introduces a time lag between the opening of the valve and the raising of the level in the reboiler. Failure of the controller or the valve will result in a low liquid level in the reboiler, which might burn out a fire tube. The overflow system with the surge tank directly below the reboiler minimizes this hazard by simply insuring the existence of a minimum liquid level. Where the fuel for the reboiler is not constant or in areas of high winds, the operators usually equip the units with a pump shutdown device in the event of flame failure. This prevents the glycol from circulating when the reboiler is out of service, which prevents the system from building up with water and minimizes glycol loss.

The still column, operating at the highest temperature in the system, shows the effects of the highest

64

rates of corrosion. Reboiler corrosion can be minimized by keeping the system free of contaminants and operating at a minimum temperature as explained elsewhere. One method to keep the temperature down is to elevate the reboiler to provide the necessary gravity head to return the glycol to storage and operate the reboiler at atmospheric pressure. Another proposed remedy is to reduce the capacity of the reboiler so that the temperature is raised and lowered quickly, which minimizes the time the glycol is at the elevated temperature and subject to decomposition. Some operators have recently gone to stainless steel construction, but usually most designs include an ample metal thickness to take care of corrosion.

Another problem frequently encountered in dehydration plant operation is foaming. Glycols foam with hydrocarbon condensate emulsion, some corrosion inhibitors, and corrosion products. Removal of these materials before they enter the absorber is the most reliable preventive measure. In some cases, defoaming agents are effective. Foaming problems require individual attention since each case is unique and no set formula can be applied.

The loss of glycol is one of the most common and consistent problems facing the operator of a dehydration plant. Some glycol losses are normal and unavoidable; the best of systems will lose about 0.1 gal per 1 million cu ft of gas passing through the absorber. The losses may be slightly higher than this at very low pressures or very high temperatures. These unavoidable losses are the small amount of glycol that vaporizes as the gas bubbles through the glycol in the absorber or that leaves in the form of mist from the top of the absorber. Another normal and unavoidable loss is that small amount of glycol that vaporizes with the water in the reconcentrator still column. It is not economical to reduce this loss beyond a certain point because of the high cost of elaborate refluxing systems. Glycol can be lost through leaks from the bottom or chimney tray into the integral scrubber section of an absorber tower. This lost glycol would be discharged with the condensate or water and would be difficult to detect. A leak in the glycol-gas heat exchanger could cause glycol losses and would be equally difficult to locate. Glycol may be lost through the absorber outlet due to a foaming condition, usually caused by condensate in the gas, compressor oils, or high concentrations of solid particles. Many effective antifoaming agents are available, but a high-quality inlet-gas scrubber is usually the best solution to the problem. Foaming

may also occur in the reconcentrator, and cause glycol losses out the still-column vent line. Plugging of the still-column packing is another cause for glycol losses; this is almost always caused by deposition of solid particles in the packed still column or by breakage and compaction of the still-column packing material.

Still-column packing, which is normally a ceramic-like material, can be broken if a slug of cold liquid hits the relatively hot packing. Repeated heat shocks will ruin the packing in time. If a slug of free water or light condensate should enter the absorber, it would soon be dumped to the reconcentrator where it would vaporize or flash and cause excessive vapor velocity through the still column. This would carry out any glycol in the column and, at the same time, could lift and drop the packing and damage it. Improper start-up procedures can cause glycol losses. The sales-gas line downstream of the absorber should be pressured up very slowly or the absorber bypassed before putting the system on stream. A low pressure in this sales line can cause excessively high gas flow rates through the absorber for a short period of time and cause all the glycol in the absorber to be blown down the line. Glycol can be lost if the trays in the absorber are not filled before start-up.

SOLID-DESICCANT DEHYDRATORS

Where the highest possible dew-point depression is required, the solid- or dry-desiccant dehydration system is the most effective type. It is not uncommon to process gas through these systems with a resultant residual water vapor in the outlet gas of less than ½ lb per MMscf. In the average system, this might correspond to a dew point of -40 F. Dehydrators of this type are manufactured as packaged units ranging in capacity from 3 to 500 MMscf/D with design pressures of from 300 to 2,500 psig. Solid-desiccant units find their greatest application in gas transmission line systems.

The essential components of a solid-desiccant dehydrator installation are—

1. an inlet-gas stream separator, usually a filter separator;
2. two or more adsorption towers (adsorbers or contactors) filled with a granular gas-drying material;
3. a high-temperature heater to provide hot regeneration gas for drying the desiccant in the towers;

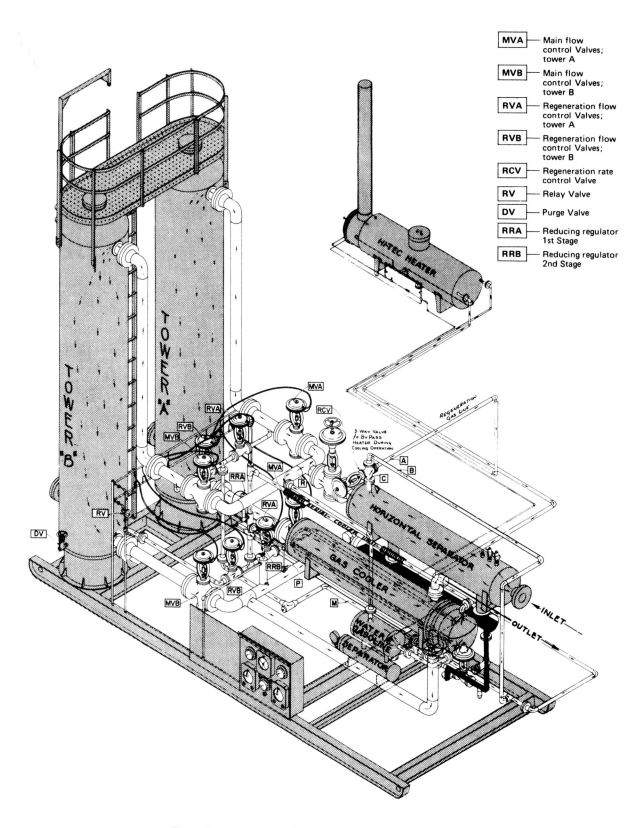

MVA	Main flow control Valves; tower A
MVB	Main flow control Valves; tower B
RVA	Regeneration flow control Valves; tower A
RVB	Regeneration flow control Valves; tower B
RCV	Regeneration rate control Valve
RV	Relay Valve
DV	Purge Valve
RRA	Reducing regulator 1st Stage
RRB	Reducing regulator 2nd Stage

Figure 6.4. Two-Tower Solid-Dessicant Dehydration Unit

66

4. a regeneration-gas cooler for condensing water from the hot regeneration gas;
5. a regeneration-gas separator to remove water from the regeneration-gas stream; and
6. piping, manifolds, switching valves, and controls to direct and control the flow of gases according to process requirements.

Figure 6.4 illustrates an installation embodying these essential components.

The following terms apply to the technology of solid-desiccant dehydrations:

Wet gas is gas containing water vapor prior to flowing through the adsorber towers.

Dry gas is gas that has been dehydrated by flowing through the adsorber towers.

Regeneration gas is wet gas that has been heated in the regeneration-gas heater to temperatures of 400 F to 460 F. This gas is passed through a saturated adsorber tower to dry the tower and remove the previously adsorbed water.

Desiccant is a solid, granulated drying or dehydrating medium that has an extremely large effective surface area per unit weight because of a multitude of microscopic pores and capillary openings.

A typical desiccant might have as much as 4 million sq ft of surface area per 1 lb.

The term *adsorption* refers to the effect that natural forces have on the surface of a solid in tending to capture and hold vapors and liquids on its surface. Adsorption processes, as opposed to absorption processes, do not involve chemical reactions. Adsorption is purely a surface phenomenon. All solids adsorb water to some extent, but their efficiency varies primarily with the nature of the material, its internal connected porosity, and its effective surface area. In most dehydration systems, activated alumina (bauxite) or a silica-gel-type desiccant is used. Adsorbents are specific in nature, and not all adsorbents are equally effective. Different molecules will be attracted to adsorbents at different rates. Because of this, adsorbents are capable of separating materials preferentially, in either gaseous or liquid phases. This is accomplished by passing the stream to be treated through the tower packed with adsorbent. The degree of adsorption is a function of operating temperature and pressure; adsorption, up to a point, increases with pressure increase and decreases with a temperature increase. A bed may be regenerated by either decreasing its pressure or by increasing

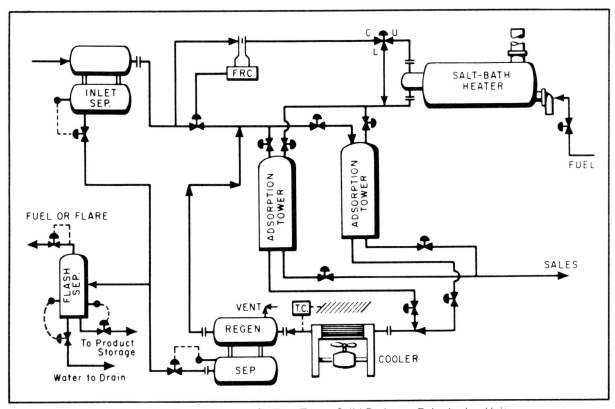

Figure 6.5. Flow Diagram of a Two-Tower Solid-Desiccant Dehydration Unit

67

Figure 6.6. Three-Tower Solid-Desiccant Dehydration Units for a Transmission Line

its temperature. For practical reasons, dehydration towers are regenerated by increasing the bed temperature by passing a stream of very hot gas through the bed. The hot natural gas not only supplies heat but also acts as a carrier to remove the water vapor from the bed. After the bed is heated to a predetermined temperature, it is cooled by the flow of unheated gas and thus made ready for another adsorption cycle.

Figure 6.5 is a flow diagram of a two-tower solid-desiccant dehydration unit. The wet-inlet-gas stream first passes through an efficient inlet separator where free liquids, entrained mist, and solid particles are removed. This is a very important part of the system since free liquids may damage or destroy the desiccant bed and solids may plug it. If the plant happens to be downstream of an amine unit or a compressor station, a filter-type inlet separator should be used. At any given time, one of the towers will be on stream in the adsorbing or drying cycle and the other tower will be in the process of being regenerated and cooled. Several automatically operated switching valves and a controller route the inlet gas and regeneration gas to the proper tower at the proper time. Typically, a tower will be on the adsorb cycle for from 4 to 12 hr, with 8 hr being the most

common time cycle. The tower being regenerated would be heated for about 6 hr and cooled during the remaining 2 hr. Large volume systems may have three towers. At any given time, one tower would be in the *adsorption cycle,* one tower would be in the heating cycle, and the remaining tower would be in the cooling cycle. Figure 6.6 illustrates several three-tower dehydration units installed for a transmission pipeline.

As the wet inlet gas flows downward through the tower on the adsorption cycle, all of the adsorbable gas components are adsorbed at different rates. The water vapor is immediately adsorbed in the top layers of the bed. Dry hydrocarbon gas components (ethane, propane, butane, etc.) passing on down through the bed are also adsorbed, with the heavier components displacing the lighter components as the cycle proceeds. As the upper layers of desiccant become saturated with water, the lower layers begin to see wet gas and begin adsorbing the water vapor, displacing the previously adsorbed hydrocarbon components. For each component in the inlet-gas stream, there will be a section of bed depth, from top to bottom, where the desiccant is saturated with that component and where the desiccant is just starting to see that component. The depth of bed from saturation to initial adsorption is known as the mass-transfer zone. This is simply that zone or section of the bed where a component is transferring its mass from the gas stream to the surface of the desiccant. As the flow of gas continues, the mass-transfer zones move downward through the bed and water displaces all the previously adsorbed gases until finally the entire bed is saturated with water vapor. When the bed is completely saturated with water vapor, the outlet gas would be just as wet as the inlet gas. Obviously, the towers must be switched from adsorb cycle to regeneration cycle before the bed has become completely saturated with water.

Regeneration gas is supplied by taking a portion of the entering wet-gas stream across a pressure reducing valve that forces a portion of the upstream gas through the regeneration system. In most plants, a flow controller regulates the volume of regeneration gas taken. This gas is sent through a heater, usually a salt-bath type, where it is heated to about 400 F to 450 F and then piped to the tower being regenerated.

Figure 6.7 illustrates the relationship between regeneration-gas temperature and desiccant-bed temperature for a typical 8-hr regeneration cycle. Initially, the hot regeneration gas must heat up the tower and the desiccant. At about 240 F, water will begin boiling or vaporizing and the bed continues to heat up but more slowly since water is being driven out of the desiccant. After all the water has been removed, heating is continued to drive off any heavier hydrocarbons and contaminants, which would not vaporize at low temperatures. With cycle times of 4 hr or greater, the bed will be properly regenerated when the outlet-gas temperature has reached 350 F to 375 F. Following the heating cycle, the bed will be cooled by flowing unheated regeneration gas through the bed. The cooling cycle will normally be terminated when bed temperature has dropped to about 125 F since further cooling might cause water to condense from the wet-gas stream and to presaturate the bed before the next adsorption cycle begins. All the regeneration gas used in the heating and cooling cycle is passed through a heat exchanger, normally an

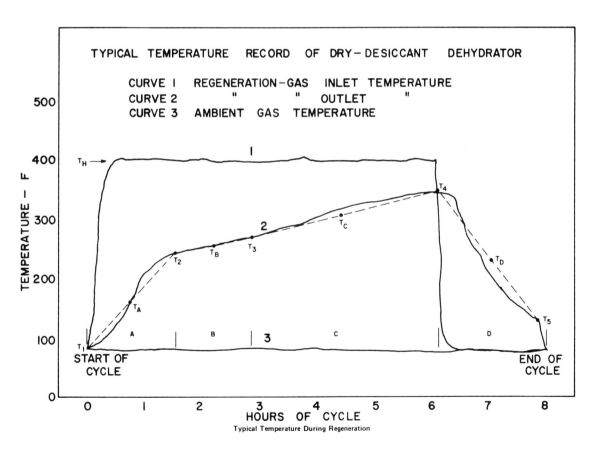

Figure 6.7. Regeneration-Gas Temperature versus Desiccant-Bed Temperature in a Dry-Desiccant Dehydrator

aerial cooler, where it is cooled in order to condense the water removed from the regenerated tower. This water is separated in the regeneration-gas separator, and the gas then is mixed with the incoming wet-gas stream. This entire procedure is continuous and automatic and, basically, is a simple and trouble-free operation.

The usable life of a desiccant may range from 1 to 4 yr in normal service. All desiccants become less effective in normal use through loss of effective surface area. The loss of effective surface area is rapid at first and then becomes more gradual as the desiccant ages. Abnormally fast degradation occurs through blockage of the small pores and capillary openings, which contain most of the effective surface area. Lubricating oils, amines, glycols, corrosion inhibitors, and other contaminants, which cannot be removed during the regeneration cycle, will eventually ruin the bed. Hydrogen sulfide may poison the desiccant and reduce its capacity. Light liquid hydrocarbons may accumulate if inadequate regeneration temperatures are not attained. Activated alumina has good resistance to liquids, but it may tend to powder due to mechanical agitation of the flowing gas.

Automatically operated dry-desiccant dehydration units are very satisfactory for continuous operation. For units that have been well designed, most of the day-to-day operating problems are minor and of a mechanical nature. However, with long-time usage of the equipment and desiccant, some more serious operating problems may begin to occur. The importance of protecting the desiccant beds from slugs of liquid water and liquid hydrocarbons should be emphasized. Some of the better desiccants now available are highly susceptible to failure when doused with liquid water. Severe fouling of the dry-desiccant bed may occur in a very short time when treating a gas stream containing both hydrogen sulfide and oxygen. Although the relative concentrations of these two components may be very small, the desiccant acts as a catalyst to convert the hydrogen sulfide to free sulfur and water. The formation of free sulfur in the pores of the desiccant plugs up the pores so that the desiccant has little capacity for removing water vapor. Corrosion is usually not a serious problem in dry-desiccant dehydration; however, where considerable carbon dioxide and hydrogen sulfide is present, corrosion may occur in the regeneration-gas heat exchanger. At this point, there is free water condensing inside the heat exchanger in the presence of these acid gases. Extreme corrosion has been noted where both carbon dioxide and

oxygen are present in the gas to be dehydrated. Occasionally corrosion in the regeneration equipment has been combated by the injection of ammonia ahead of the regeneration-gas cooler. Although this usually helps to control the pH of the water that is being removed, it has an adverse effect upon the operation of the dry desiccant. Ammonia and carbon dioxide react to form an unstable white solid that plugs up the pores of the desiccant. Upon termination of the injection of ammonia and repeated regeneration of the bed, much of the carbonate that has been formed will decompose, restoring some of the desiccant capacity for removing water. In some cases, continuous attrition of the desiccant has resulted in partial plugging of the bed and difficulty with compressors downstream. The desiccant breakage is usually the result of too high gas velocities, slugs of free water reaching the desiccant, or sudden pressure surges. It frequently has been necessary to install a special filter in the main-gas stream behind the desiccant beds.

Occasionally the main-gas temperature has been too hot immediately following the switch to the regenerated-desiccant bed. If this occurs when the gas is flowing at the full design capacity of the unit, it frequently is the result of too long a heating cycle and too short a cooling cycle. Even when the outlet-gas temperature immediately after the tower switch is satisfactory at the full design capacity, this temperature becomes higher as the unit handles less gas. This condition results from the fact that the quantity of desiccant and the volume of steel necessary to be heated in the regeneration process is constant. As the quantity of main-gas flow decreases, there is less mass for cooling the reactivation-gas stream. The main-gas outlet temperature then must necessarily increase to dissipate the heat from the desiccant bed and pressure vessel. Most heaters for dry-desiccant dehydration units utilize a high-temperature salt bath. Although somewhat more costly than a direct fired tubular heater, it has several advantages. The high-temperature bath is kept at a constant temperature so that the heat is readily available for regeneration. Tube failure is greatly reduced by decreasing the metal wall temperature and eliminating severe hot spots. Direct-fired tubular heaters for this application are susceptible to failure. When this occurs considerable damage may result since the gas in the tubes is at full line pressure.

When a unit is operated at some pressure less than that for which the installation was designed, throughput should be reduced to maintain the desired efficiency. A given volume passing through the sorber at reduced pressure can do so only by reason of increased velocity. This, in turn, causes excessive pressure drop and may result in disturbance of the desiccant bed.

Performance is sensitive to the temperature of the gas supplied to the unit. All units are designed to handle a maximum volume of gas at specified temperature and pressure. If the pressure remains constant and the inlet temperature rises, the throughput of the unit should be decreased.

HYDROCARBON RECOVERY UNITS (HRU)

Methods and equipment for recovering the largest possible quantity of liquefied petroleum products from natural gas have undergone considerable change and improvement as the demand for LPG has increased. The development of hydrocarbon recovery units is an example. Although low-temperature separation units—employing either expanding gas or mechanical refrigeration for cooling the gas—and gas plants themselves are actually hydrocarbon recovery units, this term has come to apply primarily to dry-desiccant adsorption units. These units can be designed to operate profitably on relatively small or quite large volumes—5 MMcf to 100 MMcf or more—of low-liquid-content gas, for example, 0.25 to 0.50 gal of propane and heavier components per 1,000 cu ft (GPM). The adsorption process employed in these units is a variation of the dry-desiccant dehydration process. It will be recalled that in the dehydration process, water is adsorbed first in the top of the desiccant bed and hydrocarbons, selectively heavier to lighter, are adsorbed in lower levels of the bed and are later replaced by water as the bed is saturated from top to bottom. For hydrocarbon recovery, the so-called quick-cycle system is used in which the adsorption cycle is quite short, lasting only 15 to 20 min. This is long enough so the heavier hydrocarbons can replace the methane and ethane, but short enough so the water does not replace the heavier hydrocarbons. Thus, instead of driving off only water in the drying cycle, both water and hydrocarbons are driven off. These are condensed in coolers, the water being discarded and the hydrocarbons saved. Figure 6.8 gives a flow diagram of a three-tower HRU open-cycle unit. The equipment for the adsorption process is almost exactly the same as for dehydration, except three or four contactors are needed to get adequate heating and cooling due to the short adsorption cycle. Normally, while one tower is in the adsorption cycle, one is cooling and one is heating or regenerating.

Advantages of the adsorption system over other hydrocarbon recovery systems are: (1) separate dehydration is not needed; (2) little moving machinery is required, which makes maintenance easier; and (3) high pressure is not a requirement. In all systems, stage separation or stabilization is needed for the liquid production in order to hold evaporation losses to a minimum.

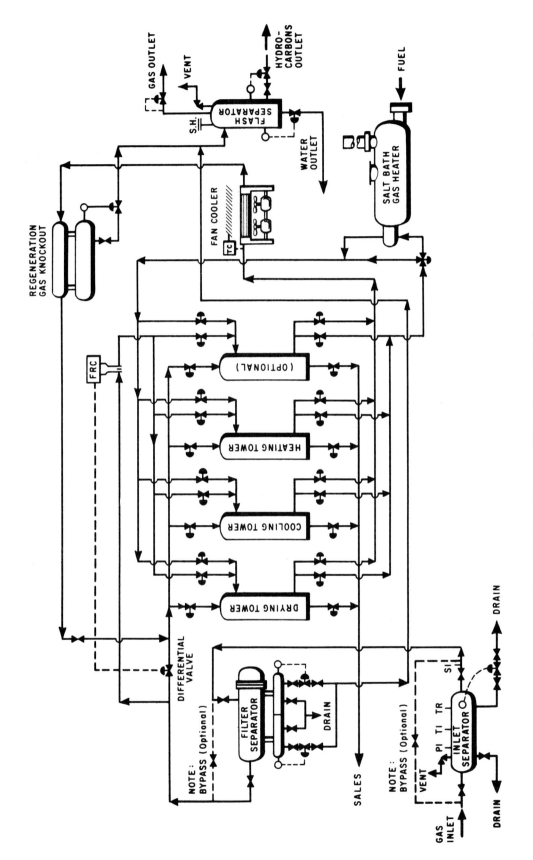

Figure 6.8. Flow Diagram of Three-Tower HRU Open Cycle

VII

MISCELLANEOUS GAS CONDITIONING

Natural gas well streams often contain hydrogen sulfide (H_2S) and carbon dioxide (CO_2). These two gases are called acid gases because in the presence of water, they form acids or acidic solutions. These gases, particularly H_2S, are very undesirable contaminants, and unless they are present in very small quantities, they must be removed from a natural gas well stream.

Most pipeline specifications limit H_2S content to 0.25 gr per 100 cu ft of gas. This is a very small amount indeed, equivalent to about four parts per million. H_2S must be removed for several reasons, the most important being that it is a toxic and very poisonous gas and cannot be tolerated in gases that may be used for domestic fuels. H_2S in the presence of water is extremely corrosive and can cause premature failure of valves, pipelines, and pressure vessels. It can also cause catalyst poisoning in refinery vessels and necessitates that many other expensive precautionary measures be taken.

The terms *sour crude* and *sour gas* mean that the crude oil or gas contains H_2S in amounts above the acceptable industry limits. The terms *sweet crude* and *sweet gas* mean either a non-H_2S-bearing oil or gas or oil or gas that has been sweetened by treating.

Carbon dioxide removal is not always required, but most treating processes that remove H_2S will also remove CO_2; therefore, the volume of CO_2 in the well stream must be added to the volume of H_2S to arrive at the total acid-gas volume to be removed. CO_2 is corrosive in the presence of water and as an inert gas has no heating value. In sufficient quantities, therefore, CO_2 might reduce the heating value (Btu/cf) below acceptable limits. Carbon dioxide removal may be required in gas going to cryogenic plants to prevent solidification of the CO_2.

Carbon disulfide (CS_2), carbonyl sulfide (COS) and mercaptans must also be considered in treating processes. Pipeline specifications normally allow 10 to 20 gr total sulfur, including the 0.25 gr of H_2S. CS_2, COS, and mercaptans must be included in the total maximum allowable sulfur content.

REMOVAL OF ACID GASES

There are several processes for removing acid gases from natural gas. Some are selective for only H_2S removal, some only for CO_2 removal. The oldest process is also the most limited. It is the iron-sponge process, which is a dry process consisting of iron oxide (Fe_2O_3) impregnated on wood chips or shavings. It is usually used on low-concentration sour gas streams whereby a vessel can operate 30 to 60 days either without any regeneration or with the partial regeneration that can be effected with air passage through the vessel. The vessel must be recharged with new iron-sponge material when gas sweetening is no longer possible.

The most widely used process in industry, the alkanolamine process, is a continuous operation liquid process using absorption for the acid-gas removal with subsequent heat addition to strip the acid-gas components from the absorbent solution. The alkanolamine absorbing solution is not selective and absorbs total acid-gas components. The process is particularly useful in order to obtain low acid-gas residual concentrations, such as are required for gas transmission pipeline gas, chemical feedstocks, and domestic household and building heating usage. The type of process involved is chemical absorption; the absorbing alkanolamine solution chemically reacts with the absorbed components.

Iron-Sponge Sweetening

The iron-sponge process is one of the oldest known for removal of sulfur compounds from industrial gases, but only in recent years has the process been applied to gas sweetening of high-pressure natural gas. The process is a batch process, the sponge being a sensitive, hydrated iron oxide (Fe_2O_3) supported on wood shavings. The reaction between the sponge and H_2S in the gas stream is—

$$2Fe_2O_3 + 6H_2S \rightarrow 2Fe_2S_3 + 6H_2O$$

The ferric oxide is present in a hydrated form, and without the water of hydration the reaction will not proceed. Thus, the operating temperature of the vessel must be kept below approximately 120 F, or a supplemental water spray must be provided.

Regeneration of the bed is sometimes accomplished by the addition of air (O_2), either continuously or by batch addition. The regeneration reaction is—

$$2Fe_2S_3 + 3O_2 \rightarrow 2Fe_2O_3 + 6S$$

Because the sulfur remains in the bed, the number of regeneration steps is limited, and eventually the bed will have to be replaced.

The iron-sponge process is most applicable for small gas volumes with low H_2S contents. It is not affected by pressure, and residue gas, well within the 0.25 gr H_2S/100 cu ft specification, can be obtained as long as the bed is not fouled. The process is selective toward H_2S. If there is CO_2 in the stream, it will not be affected. Also, for gas streams with small amounts of oxygen present, the O_2 serves to continuously regenerate the bed. The primary disadvantage of the process is the difficult changeout operation and disposal of the spent sponge.

Alkanolamine Sweetening

The term *alkanolamine* encompasses the family of specific organic compounds of monoethanolamine (MEA), diethanolamine (DEA), and triethanolamine (TEA). The chemicals are used extensively for the removal of hydrogen sulfide and/or carbon dioxide from other gases and are particularly adapted for obtaining the low acid-gas residuals that are usually specified by pipelines. The alkanolamine process is not selective and must be designed for total acid-gas removal even though CO_2 removal may not be required or desired to meet market specifications.

The process is based on the chemical reaction of a weak base (alkanolamine) and a weak acid (H_2S and/or CO_2) to give a water-soluble salt. The following typifies the reactions of acid gases with monoethanolamine.

Absorbing:

MEA + H_2S → MEA Hydrosulfide + Heat

MEA + H_2O + CO_2 → MEA Carbonate + Heat

Regenerating:

MEA Hydrosulfide + Heat → MEA + H_2S

MEA Carbonate + Heat → MEA + H_2O + CO_2

These are reversible equilibrium reactions with an increase in temperature shifting the equilibrium to the left. In general, MEA is preferable to either DEA or TEA solutions for the following reasons.

1. It is a stronger base and more reactive than either DEA or TEA.
2. It has a lower molecular weight, which means that pound for pound it requires less circulation to maintain a given amine to acid-gas mol ratio.
3. It has greater stability.
4. It can be readily reclaimed from contaminated solution by semicontinuous distillation.

MEA has the disadvantage of combining with carbonyl sulfide (COS) to form a nonregenerable compound and expending amine in this reaction. With streams containing appreciable amounts of COS, DEA is normally used, since its reaction product can be regenerated. COS is usually found in refinery cracked-gas streams but is not normally found in natural gases. However, some extremely sour natural gases do contain significant amounts of carbonyl sulfide.

The schematic flow diagram, figure 7.1, shows the basic process. The gas to be sweetened flows from its source such as wells, compressor discharge, and so forth to the inlet scrubber. This scrubber can be an integral portion of the absorber, or it can be a separate vessel, dependent upon preference and operating conditions. It is extremely important that any separable liquids be removed from the gas stream prior to the absorption process.

After removal of any separable water and/or hydrocarbons, the gas passes to the absorber section and rises countercurrently in intimate contact with the descending amine solution. Purified gas flows from the top of the absorber to the processing plant, the dehydration unit, compression, sales, or other disposition. A scrubber on the absorber outlet is very

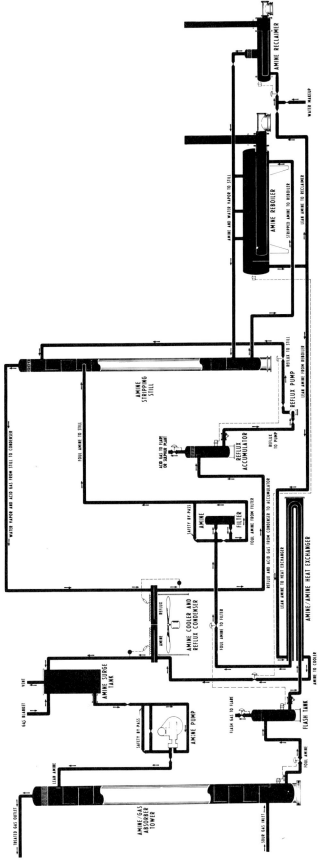

Figure 7.1. Schematic Flow Diagram of the Alkanolamine Sweetening Process

75

desirable to catch any amine solution carry-over resulting from periods of upset or unstable operation. Any amine caught in the outlet scrubber is automatically dumped back into the system via the amine surge tank.

The lean amine solution flows across the top tray and down across subsequent trays and removes acid gases from the rising gas stream. The resulting acid-gas-rich amine then leaves the bottom of the absorber through a dump valve actuated by a liquid-level controller. The rich amine then goes to the flash tank, operating at a reduced pressure, where a great portion of the physically absorbed gases are flashed off.

The rich amine then flows through the heat exchanger where it is heated by the hot lean solution coming from the still, and its temperature is increased to approximately 190 F. Then, it flows into the top of the regeneration column and is further heated by steam rising countercurrently to the descending amine solution. This heating shifts the equilibrium, which liberates H_2S and CO_2 and regenerates the MEA. The steam also serves to sweep the liberated acid gases out of the still.

Overhead vapors from the still, consisting essentially of steam and acid gases, go to the reflux condenser where the steam is condensed and the acid gases cooled. This cooling can be accomplished with water, air, or with the gas steam in some special cases.

The condenser effluent then flows to the reflux accumulator for separation of the condensed steam and acid gases. The steam condensate is returned to the still by the reflux pump. Acid gases are removed from the system through the still back-pressure valve and may go to a flare stack or a sulfur-extraction facility.

Hot regenerated amine from the still kettle flows through the heat exchanger, where it is cooled by the rich amine stream. Final cooling is accomplished by aerial coolers or by water coolers, if available. It then flows to the surge tank with flow controlled by a liquid-level controller and diaphragm valve to maintain a constant level in the reboiler. Amine pumps take suction from the surge tank and discharge to the top tray of the absorber to complete the cycle.

Heat is added to the still reboiler by steam, hot oil, or direct firing. To reduce water losses, the steam system may be a closed-circuit integrated facility with gravity return of condensate to the steam generator. It may also be existing plant steam for turbines, boilers, and so forth. Hot oil heating is accomplished by heating an oil-transfer medium in a direct-fired heater and pumping the oil through a reboiler-tube bundle. The direct-fired method of heating simply consists of controlled combustion on the inside of an expandable tube that is surrounded by the boiling amine solution.

Glycol/Amine Process

The glycol/amine process uses a solution comprised of 10 to 30 weight percent MEA, 45 to 85 percent glycol, and 5 to 25 percent water for the simultaneous removal of water vapor, H_2S, and CO_2 from gas streams. The combination dehydration and sweetening unit results in lower equipment cost than would be required with the standard MEA unit followed by a separate glycol dehydrator. Obtaining pipeline specification for H_2S in the residue gas stream does not appear to be a problem with glycol/amine since the solution, with the high boiling point glycol present, can be stripped at atmospheric pressure and fairly high temperatures. The same is not true in conventional MEA units since the high temperatures needed for more complete stripping adversely affect the MEA solution and lead to excessive degradation and corrosion problems. The degree of dehydration that can be obtained from the glycol/amine process will be as good or slightly better than that from a glycol solution with an equal amount of water, that is, a 95 percent DEG and 5 percent H_2O mixture. This should not be confused with the dehydration that can be obtained from a standard glycol dehydrator, which produces glycol with purities in excess of 99 percent.

The main disadvantages of glycol/amine process are as follows.
1. Increased vaporization losses of MEA are due to the higher regeneration temperatures.
2. Reclaiming must be by vacuum distillation.
3. Intricate corrosion problems are present in operating units, and the solution of one plant may not apply to any other plant.
4. Application must be for gas streams that do not require low dew points.

The process flow scheme is essentially the same as the MEA process.

Sulfinol Process

The Shell sulfinol process is unique in that it uses a mixture of solvents, which allows it to behave as both a chemical and physical solvent process. The solvent is composed of sulfolane, DIPA (di-isopropanolamine), and water. The relative amounts of each ingredient are varied to yield a solvent composition that is tailored to fit each application. The sulfolane acts as the physical solvent; whereas, DIPA acts as the chemical solvent. This combination of absorption capabilities offers advantages both for loading and unloading of the solvent. The sulfinol solvent has a good affinity for sour components at low to medium partial pressures and an extremely high affinity for sour components at high partial pressures.

Sulfinol appears to have its greatest advantage when the acid-gas partial pressure is about 30 psia or greater. In general, COS, CS_2, and mercaptans can be satisfactorily removed from the feed gas, along with H_2S and CO_2, within certain limitations of concentrations in the feed gas. Degradation of sulfinol solvent by COS, CS_2, and H_2S is practically nil. CO_2 gradually degrades the DIPA to DIPA-Oxazolidone, which can be readily removed from the system by a simple reclaimer unit.

The main advantages of sulfinol are: (1) low solvent circulation rates; (2) smaller plant equipment and lower plant cost; (3) low heat capacity of the solvent; (4) low utility costs; (5) low degradation rates; (6) low corrosion rates; (7) low foaming tendency; (8) high effectiveness for removal of COS, CS_2, and mercaptans; (9) low vaporization losses of the solvent; (10) low heat-exchanger fouling tendency; and (11) nonexpansion of the solvent when it freezes.

Disadvantages of sulfinol are: (1) the absorption of heavy hydrocarbons and aromatics and (2) the expensive nature of the sulfolane in the solvent. Sulfinol is a proprietary process and requires payment of a royalty to Shell for its use.

Molecular-Sieve Removal of H_2S and CO_2

In addition to the amine-solution and sulfinol-solution processes, H_2S and CO_2 can be removed by the use of mol sieves in a dry-bed unit such as

discussed previously. Today the major difficulty with these units is how to dispose of the sour regeneration gas. Usually this gas is flared or burned in an incinerator. With the emphasis on environmental protection today, operators are finding it more and more difficult to vent or burn any plant effluents such as these. Some consideration has been given to using a small MEA system in combination with a sulfur plant to handle these effluents.

BTU CONTROL

As more and more ethane and propane are extracted from natural gas, the control of the Btu or heating content of the gas becomes more important. Small, low-temperature gasoline plants are being installed, which are capable of removing all the propane and heavier components and 60 to 70 percent of the ethane. The residue gas from such plants often consists of little more than methane plus inert gases such as CO_2 or N_2. If the feed gas contains substantial quantities of such gases, the heating value of the residue gas may fall below contract specifications (usually in the order of 1,000 Btu/cf). Thus, the amount of ethane or propane extracted may be limited to something less than that economically possible otherwise.

When possible, low-Btu gases are blended with gases having fairly high Btu values so that the average meets specifications. Such blending requires a high degree of care and precision, especially if the gas is to be used for residential heating. Variations in the gas composition will cause smoking or flameouts because the air-mixture setting on burners is fixed. In most cases where blending takes place or plant operations can cause a variation in the Btu content, recording calorimeters are used to monitor the residue-gas stream. Adjustments are made in the blending or plant operation to maintain constant a Btu content based on data obtained from the calorimeter.

VIII

COMPRESSORS AND PRIME MOVERS

A most common item of equipment used in the handling and transporting of gas is the compressor. Compressors may vary in size from small belt-driven units in the order of 50 hp to very large reciprocating or centrifugal units of 15,000 hp or more. The most common type of compressor is the gas-engine-driven reciprocating unit. However, gas-turbine-driven centrifugal units are coming into use more and more. Other types of drivers for reciprocating units are electric motors and steam turbines. These drivers are ordinarily used in special situations and are rarely found in field operations. Electric motors are normally difficult to justify since electric power usually costs more than the fuel gas that would be saved. Expansion turbines are used quite often in plants or refineries where a pressure drop in a fluid stream (gas, hot oil, etc.) can be utilized.

RECIPROCATING COMPRESSORS

Drivers

The internal-combustion gas engine is the leading prime mover for gas-compression service in the oil industry. The popularity of the gas engine in this field has been maintained primarily because of its overall efficiency, because of the availability of a clean, relatively low-cost fuel, and because manufacturers of this equipment have made gas engines and reciprocating compressors as compact integral units. Horsepower ratings and general design of modern gas engines are significantly different from the units manufactured a generation ago. Today, integral gas-engine-compressor units are available from under 200 hp to the extremely large sizes of 7,500 hp or more. Guaranteed thermal efficiencies of some of the integral gas-engine-compressor units are now approaching 40 percent. For comparison, maximum thermal efficiency of a stream turbine is approximately 33 percent. Gas-turbine-engine thermal efficiency may be less than 25 percent, depending to a great extent on the utilization of the heat in the exhaust gases.

Gas engines may be classified in a number of ways including the following.

1. Combustion cycle
   a. Four-stroke cycle
   b. Two-stroke cycle
2. Power impulse
   a. Single acting
   b. Double acting
3. Cylinder arrangement
   a. Vertical
   b. Horizontal
   c. V-type
   d. Opposed
   e. Radial
4. Speed
   a. Low (100 to 250 rpm)
   b. Intermediate (250 to 600 rpm)
   c. High (600 to 1,200 rpm)

In general, the gas engines most commonly used now are both two- and four-stroke cycle units with single-acting power cylinders in a vertical or V-type arrangement operating in the intermediate speed range. The smaller units connected to compressors by belts are all considered to be high-speed units. Direct-drive units or those in which the engine crankshaft is connected to the compressor crankshaft with a coupling or clutch usually operate at high speeds. Integral units, those which provide for mounting of compressor cylinders on the engine frame, fall in the low- and intermediate-speed category. All of the various types of gas engines are basically reciprocating machines with rotating crankshafts, but perhaps the most fundamental difference is in the combustion cycle. Neither four- or two-stroke types have any clear-cut overall advantage; years of competition among engine builders have failed to eliminate either type. Figure 8.1 depicts a belt-driven single-stage compressor unit with a 160 hp multicylinder engine as driver. Figure 8.2 illustrates a typical packaged direct-drive horizontally opposed compressor arrangement. Figure 8.3 shows an integral engine-compressor unit installation.

The conventional two-stroke cycle engine consists of a combination gas engine and scavenging cylinder (air compressor) built into one compact unit. (See fig. 8.4A.) Two-stroke cycle gas engines require two piston strokes and one full revolution for each cycle. The exhaust ports in the cylinder walls are uncovered by the piston near the end of the expansion stroke permitting the escape of exhaust gases and reducing the pressure in the cylinder. The charge of air flows into a scavenger cylinder and is compressed to a few pounds per square inch above atmospheric pressure. Intake ports are uncovered by the piston soon after the opening of the exhaust ports, and the compressed charge of air flows into the cylinder expelling most of the exhaust products. (See fig. 8.4B.) When the

piston starts toward the cylinder head, it closes off the intake ports and then the exhaust ports. The fuel gas is injected into the cylinder through a valve in the cylinder head and is mixed with the supply of air that has just been admitted. As the piston moves toward the cylinder head, the compression of the fuel gas and air mixture continues until ignition, after which the piston is forced back as the charge expands. As the piston approaches the end of its power stroke, the exhaust ports are uncovered, as mentioned above, completing the cycle. Construction of a typical two-cycle engine is shown in figure 8.5.

The schematic operation of a conventional four-cycle engine is shown in figure 8.6. The four-stroke cycle engine requires four piston strokes and two complete revolutions to complete each cycle. The intake and exhaust valves are closed when the piston is near the end of its compression stroke and ignition occurs. The explosion forces the piston away from the cylinder head. When the power stroke is ended, the piston moves toward the cylinder head, the exhaust valves open, and the exhaust products are expelled. As the piston starts toward the crankshaft for the second time, the intake valve opens and the fuel-air mixture is drawn into the cylinder. The piston then starts toward the cylinder head again on the compression stroke. Ignition occurs when the piston nears the head end of the cylinder, which is at the top

Figure 8.1. Belt-Driver 160 HP Single-Stage Reciprocating Compressor

Figure 8.2. Packaged Direct-Drive Engine-Compressor Unit

center of the crankshaft revolution, and the cycle is completed. It will be noted that four strokes are required to complete this cycle: (1) intake, (2) compression, (3) power, and (4) exhaust. Construction of a typical four-cycle engine is shown in figure 8.7.

The following generalizations can be made about two- and four-stroke cycle gas engines.

1. The fuel consumption of slow-speed integral-type units varies between 9,000 and 7,000 Btu/bhp/hr depending upon the type, size, and make with the lower fuel consumption resulting from turbocharging.

2. Maintenance costs are approximately the same for the two types. Generally speaking, however, as the size of the engine increases, the unit costs in terms of dollars per brake horsepower per year ($/bhp/yr) will gradually decrease.

Horsepower ratings of engines depend mainly upon the amount of air that can be supplied to the power cylinders. Because a very definite air-fuel ratio is required, an important factor in the design of engines is the method of supplying the air for combustion. In naturally aspirated four-stroke cycle

Figure 8.3. Integral-Drive Engine-Compressor Unit

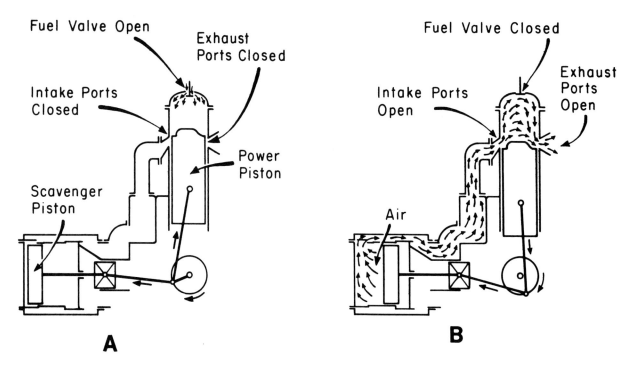

Figure 8.4. Schematic Sketch of a Two-Stroke Cycle Engine Operation

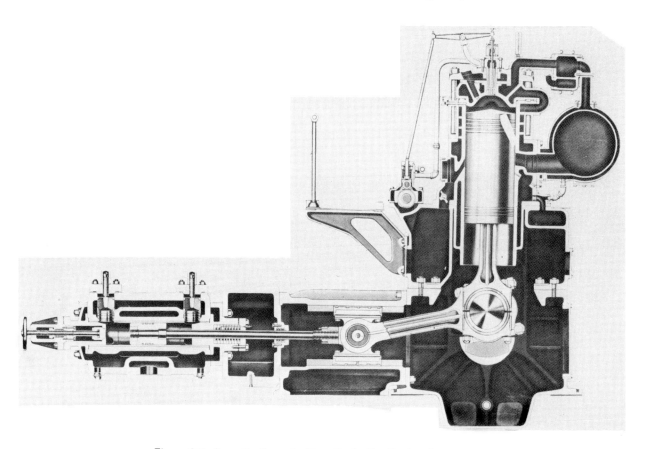

Figure 8.5. Cross Section of a Two-Cycle Gas-Engine Compressor

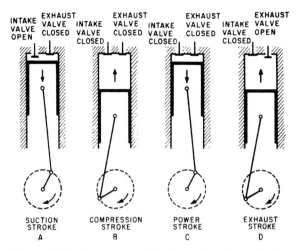

SUCTION STROKE A   COMPRESSION STROKE B   POWER STROKE C   EXHAUST STROKE D

Figure 8.6. Schematic Sketch of a Four-Stroke Cycle-Engine Operation

engines, the vacuum created by the piston moving downward sucks air into the cylinder when the intake valve is open. Conventional two-stroke cycle engines, of course, have reciprocating air-compressor cylinders built integrally into the unit to provide the necessary air for scavenging and combustion. In recent years, attention has focused on ways and means of providing additional air so that additional fuel may be burned in order that the horsepower rating of an engine can be increased without altering the engine frame. The injection of additional air into the power cylinders so more fuel can be used is called *supercharging*.

Supercharging an engine may be done in several ways.

1. Using additional air-scavenging cylinders, which is possible only if space for an additional cylinder is available and for two-stroke cycle engines only
2. Using larger air-scavenging cylinders, which is possible for two-stroke cycle engines only
3. Using air blowers driven by belts off the engine flywheel
4. Using an electrically driven blower
5. Using an exhaust gas-turbine-driven centrifugal compressor

All of the above methods have been tried successfully by various companies. In general, method no. 5 is used only on new units. A cross section of a typical exhaust gas-turbine-driven centrifugal compressor (blower) is shown in figure 8.8. Supercharging of old units should be done only after consultation with the manufacturer to determine that the bearings, frame, and crankshaft can satisfactorily handle the increased mechanical load.

Compressors

The vast majority of compressors purchased for use in gas operations are the reciprocating piston type. Basically this type of compressor consists of a ringed piston, cylinder, cylinder head, and suction and discharge valves, plus the mechanism necessary to convert rotary motion to reciprocating motion—that is, the connecting rod, the crosshead, the wrist pin, and the piston rod as shown in figure 8.9. Valves are of the automatic plate type, which depend on differential pressure for opening and closing. (See fig. 8.10.)

The compressor cylinder itself is fabricated of cast iron, nodular iron, cast steel, or forged steel. These materials are chosen in order to achieve successively higher maximum working pressures. Depending to a large extent on cylinder size, the cast-iron cylinders, which are generally the most economical, will be good for working pressures as high as 1,500 psi. In the larger sizes, of course, the working pressure of cast-iron cylinders must be reduced drastically, and in such cases nodular iron may be substituted to increase the working pressure. Nodular iron, sometimes also called spheroidal or ductile iron, is in reality a high-grade cast iron in which the graphite present has been arranged in substantially spherical shape rather than in the usual flake form as in gray cast iron. The relative tensile strengths are roughly 45,000 psi to 67,000 psi. Cast-steel cylinders are usually rated to working pressures of 2,500 or 3,000 psi, and forged-steel cylinders are used for higher pressures.

Reciprocating compressor cylinders may be purchased in either solid body or liner type. Solid-body-type cylinders are limited to one diameter and can be altered in size only by boring out the cylinder. Such altering can be done only within narrow limits because the cylinder working pressure is reduced by the change. Liner-type cylinders are more flexible because a machined liner is installed inside the body of the cylinder to provide the desired bore. Should the need for a different size cylinder occur, the liner can be removed and a different size liner and piston fitted into the body of the cylinder. Naturally, there are limitations as to the magnitude of diameter variation in each direction.

Reciprocating compressors are suitable for low or high ratios and varying rates and pressure conditions; can be built to handle small, intermediate, or high

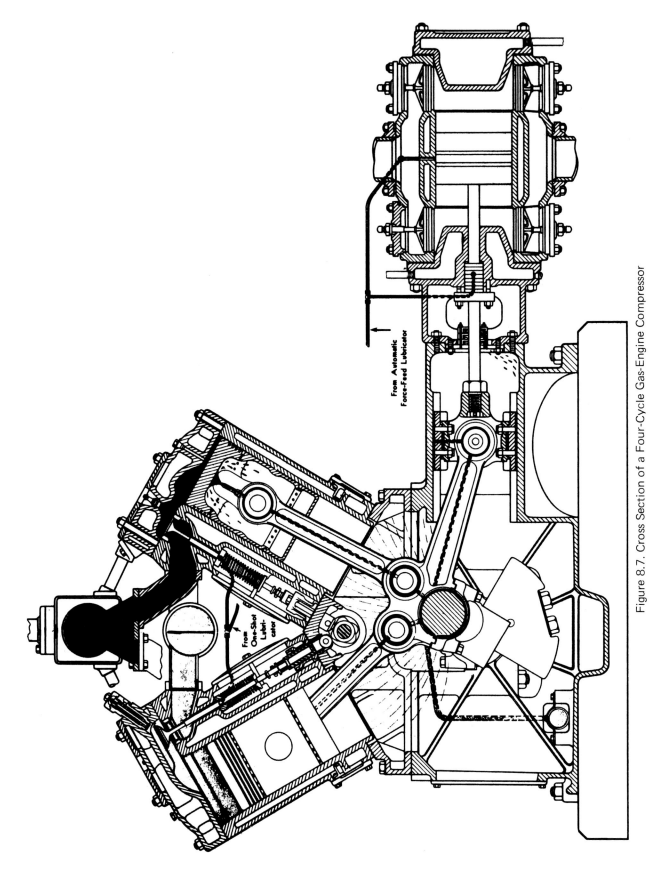

From Automatic
Force-Feed Lubricator

From
One-Shot
Lubricator

Figure 8.7. Cross Section of a Four-Cycle Gas-Engine Compressor

84

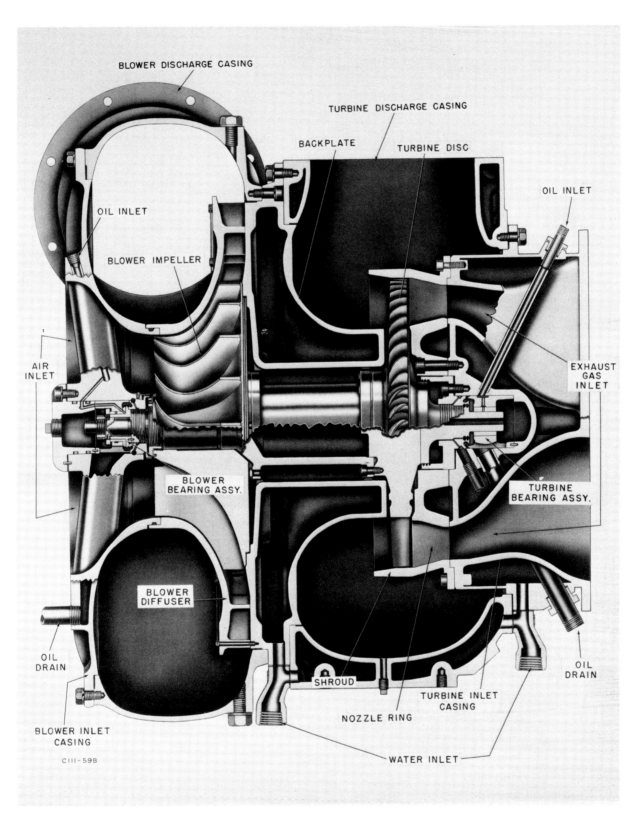

BLOWER DISCHARGE CASING

TURBINE DISCHARGE CASING

BACKPLATE

TURBINE DISC

OIL INLET

OIL INLET

BLOWER IMPELLER

AIR
INLET

EXHAUST
GAS
INLET

BLOWER
BEARING ASSY.

TURBINE
BEARING ASSY.

BLOWER
DIFFUSER

OIL
DRAIN

SHROUD

OIL
DRAIN

BLOWER INLET
CASING

TURBINE INLET
CASING

NOZZLE RING

WATER INLET

CIII-59B

Figure 8.8. Cross Section of an Exhaust-Gas Turbine-Driven Centrifugal Compressor (Blower) for Supercharging

85

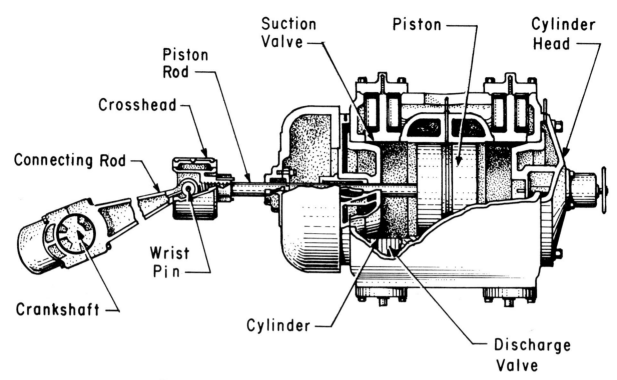

Figure 8.9. Elements of a Typical Reciprocating Compressor

throughput volumes; and are therefore most suitable to gas-production operations.

RECIPROCATING COMPRESSOR OPERATIONS

Theoretical Considerations

Gas laws are commonly considered to be rules or laws of nature that govern the behavior of gases under specified conditions. As discussed previously, gases react to pressure and temperature according to well-defined rules. These laws are directly applicable to the handling and processing of natural gas.

Work is performed when a force acts through a distance or the resistance to motion is overcome. No work is done unless motion is produced. Although gas molecules are constantly in motion overcoming internal resistance and causing pressure in a closed container, work is performed only when this pressure or force causes movement through expansion, such as pushing on a moving piston. *Force* is that which causes, changes, or stops the motion of a body. Mathematically, work is force times distance. No one has ever seen a force. Only its effect may be seen or

felt. The effect of centrifugal force is felt when turning a corner, but the force is not seen.

Energy is the capability of a body for doing work. Potential energy is this capability due to the position or state of the body. Kinetic energy is the capability due to the motion of the body. A pendulum, swinging back and forth, constantly changing its motion is simply a trading of potential and kinetic energy. Gas molecules in motion are doing the same thing that the pendulum is doing. The amount of work actually done depends upon the amount of energy available. If an engine is capable of using more gas for fuel and thus doing more work, it is capable of just that much more energy. Molecules of gas may be moving at a certain rate and thus produce a certain pressure, but if an increase in temperature will cause them to move twice as rapidly, then their energy and resulting pressure is increased.

Power is the rate of doing work or the amount of work done in a specific unit of time. It is calculated in foot-pounds per minute. One horsepower equals 33,000 foot-pounds per minute.

Velocity is the speed of a body in motion and is measured in distance per unit of time, such as feet per second. Gas molecules are minute bodies of matter

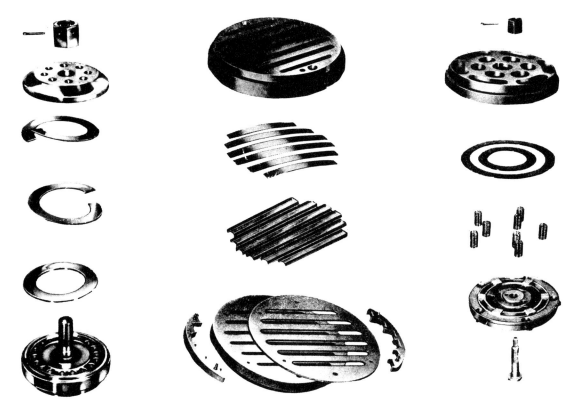

Figure 8.10. Several Types of Compressor Valves

that are always in motion except at absolute zero temperature. Their velocity determines the pressure a gas exerts and has a bearing on the work required to process, compress, or cool the gas.

Isothermal compression of air or gas exists when the interchange of heat between the air or gas and surrounding bodies (i.e., cylinders or pistons) takes place at a rate exactly sufficient to maintain the air or gas at constant temperature as the pressure increases. *Adiabatic compression* of air or gas exists when no heat is transferred between the air or gas and surrounding bodies (as the cylinders and pistons in a compressor). It is characterized by an increase in temperature during compression and a decrease in temperature during expansion. No compressor operates under either condition perfectly, but its operation may be between the two. Some conditions may make one compressor operate near isothermal and another near the adiabatic compression cycle.

Specific heat, expressed in Btu, is the amount of heat required to raise one pound of gas one degree F. The amount varies depending upon whether pressure or volume is held constant. If the pressure is maintained constant, it is called the specific heat at constant pressure and is denoted by C_p. If the pressure varies and the volume is maintained constant, it is called the specific heat at constant volume and is denoted by C_v. The specific heat also will vary with the temperature for all except the monatomic gases, which have only one atom in each molecule (e.g., helium, neon, krypton, xenon, and mercury vapor). The specific heat of other gases, such as oxygen, hydrogen, nitrogen, carbon dioxide, steam, and methane, are dependent upon the temperature range through which calculations are to be made. The operator must depend upon tables of collected data and calculations from actual compressor operation for their behavior. The ratio between the specific heat at constant pressure and the specific heat at constant volume is known as the N or K value of the gas. For example, it has been experimentally found that it takes 0.2375 Btu to raise 1 lb of dry air at 60 F, 1 F at constant pressure. If the volume had remained constant and the pressure had been allowed to increase, it would have required 0.1689 Btu. The N value in this case would be—

$$N = \frac{C_p}{C_v} = \frac{0.2375}{0.1689} \text{ or } 1.406$$

The N value depends upon the specific property of the gas. The N value for air is 1.406, approximately 1.265 for dry natural gas, and 1.1 for wet casing-head gas. Since there is such a wide variation in wetness and N value of natural gas, a curve has been prepared that gives the N value as a function of the molecular weight or specific gravity of the particular gas. If a fractional analysis is available for the gas, the molecular weight may be calculated and used to determine the N value. If the molecular weight is not available, the specific gravity can be determined and used. The power required to compress a given volume of gas is affected appreciably by the N value of the gas.

Compression Ratio, Clearance, and Volumetric Efficiency

The *ratio of compression* of a compressor is the ratio of the absolute discharge pressure to the absolute suction pressure. For instance, a compressor with atmospheric intake at 0 psig (14.7 psia) and discharge at 40 psi gauge (54.7 psi absolute) has a ratio of compression of 54.7 to 14.7; that is, 3.72 to 1. This must not be confused with the compression ratio used in power cylinders of internal-combustion engines, which is the displacement plus the clearance divided by the clearance.

The ratio of compression for gas compressors is generally limited to less than 5.5 usually falling between 2.5 and 5.0. The reason for this is evident if the temperature and volumetric efficiency of the compressor are considered. When compression is necessary over a larger number of compressions than this, two or more stages of compression are used. To illustrate this, the following example is given. Assume the discharge pressure is 4,000 psig and the suction pressure is 500 psig. The overall compression ratio is:

$$\frac{4000 + 14.7}{500 + 14.7} = \frac{4014.7}{514.7} = 7.80:1$$

Inasmuch as it is impractical to compress through 7.80 ratios in a single stage because of the heat of compression, two-stage compression will be required, and the ratio per stage will be the square root of 7.80 or 2.79. A comparable situation could be possible with low pressures, that is, suction pressure of 15 psig with a discharge of 225 psig.

The volume delivered, regardless of the compressor stroke, will be affected by the clearance of the unit. As the piston moves back and forth within its cylinder, it must never reach the end of the cylinder or it would damage itself and the cylinder head. The

space provided for this protection plus that which exists around the valves, which are located near the cylinder ends, make up the volume known as the clearance. The clearance volume is usually expressed as a percent of piston displacement by using the following formula:

$$\text{Percent Clearance} = \frac{\text{Clearance Volume, cu in.}}{\text{Piston Displacement, cu in.}} \times 100$$

Percent clearance for double-acting cylinders is based on total clearance volume in cubic inches for both head and crank end and total piston displacement for both head and crank end. Often the clearance in cubic inches is assumed to be equal between the head and crank ends even though the crank-end displacement is less than the head-end displacement because of the rod displacement volume. For small diameter cylinders, this difference in clearance volume can be appreciable and must be taken into account in calculations.

Volumetric efficiency, E_v, represents the efficiency of a compressor cylinder to compress gas. It may be defined as the ratio of the volume of gas actually delivered, corrected to suction temperature and pressure, to the piston displacement. The principal reasons that the cylinder will not deliver the piston displacement capacity are: (1) wiredrawing, which is a throttling effect at the valves; (2) heating of the gas during admission to the cylinder; (3) leakage past valves and piston rings; and (4) reexpansion of the gas trapped in the clearance-volume space from the previous stroke. Reexpansion has by far the greatest effect on volumetric efficiency. Without going into the details of the derivation of the volumetric efficiency, which may be obtained in any thermodynamics text, the percent volumetric efficiency for natural gas may be calculated using one of the following equations:

$$E_v = \left(\frac{100 - R}{100}\right) - CI\left(R^{1/N} - 1\right)$$

$$\text{or} \quad E_v = 0.97 - CI\left[\left(R^{1/N} \times \frac{Zs}{Zd}\right) - 1\right]$$

Where

E_v	=	Volumetric Efficiency, Expressed as a Decimal (for example, .8)
R	=	Compression Ratio
CI	=	Clearance, Expressed as a Decimal
N	=	Ratio of Specific Heats
Z_s	=	Compressibility Factor at Suction P and T
Z_d	=	Compressibility Factor at Discharge P and T

The terms $(100 - R)/100$ and 0.97 will obviously be equal at a compression ratio of 3.0. This factor in the equation provides for the losses due to wiredrawing, heating of the gas, and leakage, which are the first three items mentioned above. The two equations are given because there is a disagreement between manufacturers as to the proper approach to be used since the effect of these three items cannot be measured. The volumetric efficiency obtained by either will be approximately the same over the range of compression ratios normally used in gas operations.

Cylinder Capacities

Solving compressor problems can be reduced to a relatively simple sequence of calculations by applying a few basic equations and utilizing data taken from curves contained in the NGPSA *Engineering Data Book.* However, it is always desirable to know about the origin of the basic terms, and for this reason, the section below has been included. Figure 8.11 shows a typical ideal pressure-volume diagram for a compressor cylinder with corresponding compressor piston locations during reciprocation.

Position 1. This is the start of the compression stroke. The cylinder has a full charge of gas at suction pressure. As the piston moves toward position 2, the gas is compressed along the line 1 to 2.

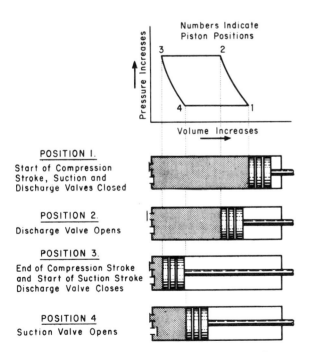

Figure 8.11. Pressure-Volume Diagram and Piston Locations During a Reciprocating Compressor Stroke

Position 2. At this point, the pressure in the cylinder has become greater than the pressure in the discharge line which causes the discharge valve to open and allows the original charge of gas to enter the discharge line. This action occurs along the line 2 to 3.

Position 3. At this point, the piston has completed its discharge stroke, and as soon as it starts its return stroke, the pressure in the cylinder drops, which closes the discharge valve. The gas trapped in the cylinder clearance volume is never discharged but expands along the line 3 to 4.

Position 4. At this point, the pressure in the cylinder has dropped below the suction pressure, which causes the suction valve to open. This permits a new charge of gas to enter the cylinder along the line 4 to 1.

Piston displacement is the actual volume displaced as the piston moves from position 1 to position 3. This factor is normally expressed in cubic feet per minute (cfm) and may be calculated as follows:

Single-acting Cylinder (Compression on One End Only)

$$PD_{sa} = \frac{A_{HE} \times S \times rpm}{1728}$$

Where

A_{HE} = Area Head End of Piston, sq in.

S = Piston Stroke, in.

rpm = Revolutions per Minute

PD_{sa} = Single-acting Piston Displacement, cfm

Double-acting Cylinder (Compression on Both Ends)

$$PD_{da} = \frac{A_{HE} \times S \times rpm}{1728} + \frac{A_{CE} \times S \times rpm}{1728}$$

$$A_{CE} = A_{HE} - A_R$$

$$PD_{da} = \frac{S \times rpm}{1728} \times (2A_H - A_R)$$

Where

A_{CE} = Area Crank End, sq in.

A_R = Area Rod, sq in.

PD_{da} = Double-acting Piston Displacement, cfm

Cylinder capacity for routine calculations can be determined using the following equation:

$$Q = PD \times E_v \times \frac{P_s}{P_b} \times \frac{T_b}{T_s} \times \frac{1,440}{1,000}$$

89

Where

Q = Cylinder Capacity, Mcf/D at Standard Conditions

PD = Piston Displacement, cfm

E_v = Volumetric Efficiency, Expressed as a Decimal

P_s = Suction Pressure, psia

P_b = Pressure Base, psia (14.7)

T_b = Temperature Base, degrees R (520 R)

T_s = Suction Temperature, degrees R

$\dfrac{1,440}{1,000}$ = Constants to Convert from cfm to Mcf/D

The definition of each of these factors has previously been given. For routine calculations, it is assumed that correction for supercompressibility is unnecessary. Where this correction is desired, it should be applied as follows:

$$Q = PD \times E_v \times \frac{P_s}{P_b} \times \frac{T_b}{T_s} \times \frac{1}{Z_s} \times \frac{1,440}{1,000}$$

Rod load is known by a variety of names including rod load, pin load, and frame load. It is possible that this confusion of names has resulted from the fact that the weak link in a compressor cylinder design is not always the same piece of equipment. On occasion it may be the rod, it may be the crosshead pin or bushing or the compressor frame. In any event, manufacturers of compressor cylinders will furnish this limit. Quite frequently the limit will be different in compression and tension and will usually be lower for single-acting than double-acting cylinders. Rod loads may be calculated as follows:

$$R_c = A_{HE} \times P_d - A_{CE} \times P_s$$

$$R_t = A_{CE} \times P_d - A_{HE} \times P_s$$

Where

R_c = Compression Rod Load, lb

R_t = Tension Rod Load, lb

A_{HE} = Area Head End, sq in.

A_{CE} = Area Crank End, sq in.

P_d = Discharge Pressure, psig

P_s = Suction Pressure, psig

The area of the crank end can be calculated by subtracting the cross-sectional area of the rod from the head-end area.

Discharge temperature after adiabatic compression may be calculated by using the following equation:

$$T_d = T_s R^{(N-1)/N}$$

Where

T_d = Discharge Temperature, Degrees R

T_s = Suction Temperature, Degrees R

R = Compression Ratio

N = Ratio of Specific Heats

Utilization of Horsepower

One of the most important factors in the operation of a gas compressor is the efficient utilization of gas-engine horsepower. By making maximum use of the horsepower available in a compressor unit, more gas can be compressed per unit. This will result in either handling more gas or operating with fewer units, either of which is economically desirable.

There are a number of factors that affect compressor unit capacity and developed horsepower, such as: (1) compressor cylinder clearance, (2) suction pressure, (3) suction temperature, (4) discharge pressure, and (5) speed. Compressors are designed originally to utilize the engine horsepower as fully as possible for the specific conditions of suction pressure, temperature, and discharge pressure under which they are to operate. Often these conditions will change from day to day and most certainly during the life of a project. For these reasons, a certain amount of flexibility in the compressor is important.

The potential horsepower built into a compressor is fixed. The engine is designed to run at a certain speed most efficiently from the standpoints of fuel consumption and maintenance. Some speed variation is possible but usually not over a very wide range. Operating an engine at a greatly reduced speed is the equivalent of purchasing a greatly oversized piece of equipment, which is poor economy. Speeding up an engine increases its horsepower output and is one way an increase in load can be handled. This can be done safely and economically only if design load limits are not exceeded and if maintenance costs do not increase too much.

The most effective way to utilize available engine horsepower when operating conditions change is to alter the effective compressor size. The most obvious and the most expensive way to do this is to install new compressor cylinders. Other ways include the changing of cylinder displacement by boring out the cylinders or changing the liner size. One of the most common approaches is to change the compressor capacity by altering the volumetric efficiency through changes in clearance. It is common practice to provide flexibility for expected load variations by this means.

A number of methods are employed for varying the clearance in compressor cylinders. Naturally, there is a certain normal clearance built into every standard cylinder fabricated by the manufacturers. To make this cylinder fit a specific case, however, the manufacturer may alter (increase) this normal clearance to what is called a built-in clearance condition. Altering (increasing) the normal clearance is usually accomplished by one or more of the following methods.

1. Remove a small portion of the end(s) of the compressor piston.
2. Add a spacer ring between the cylinder head and the cylinder body.
3. Shorten the projection of the cylinder head into the cylinder. If the head is water cooled, the manufacturer must be consulted to determine if sufficient metal is available.
4. Raise the suction and discharge valves by installing spacer rings under the valves.

After these changes have been made, to all intents and purposes the compressor cylinder has a new and larger inherent clearance.

Beside this basic change, there are other methods of varying cylinder clearance which are used very frequently. Different manufacturers utilize different procedures, but the various ways may be listed and described briefly as follows:

Installation of Head-End Fixed-Volume Clearance Pockets or "Unloaders." This method is by far the most popular and practical. A cross section of a typical head-end pocket or unloader valve is shown in figure 8.12. Essentially, this means of adding clearance is nothing more than a valve and fixed-volume chamber that can be opened or closed by a hand-wheel from outside the cylinder. Depending upon the clearance requirements as well as physical size, there may be more than one head-end fixed-volume clearance pocket built into the head-end head. Clearance pockets with variable volume are also available, but these are used less frequently because they are more expensive, maintenance costs are higher, and there appears to be little comparative advantage in their use. (See fig. 8.13.)

Side-Passage Plugs. This procedure is favored by some manufacturers and is essentially a flange with a plug attached that can be inserted into a passage built into the side of a cylinder. The number of such passages as well as the size, length, and location is dependent upon the cylinder design. For some cylinders, these side passages may be located in the head-end head of the cylinder. To provide additional clearance, the plug or some part of it, as required, is removed from the passage. This method of adding clearance is shown in figure 8.14.

External Clearance Bottles. A connection is usually provided in head-end clearance pockets to attach an external clearance bottle that provides additional clearance volume when the handwheel is opened over and above that in the pocket. For the side passages, a flanged bottle is substituted for the flange and plug discussed above when clearance volume in excess of that included in the side passages is required. There is a practical limit to how much additional clearance can be added in this manner because of the inability of the gas to squeeze in and out through the connecting passage at each stroke. One manufacturer has developed an empirical rule that limits the passage velocity to approximately 2,500 ft/min.

Crank-End (or Head-End) Valve-Cap Clearance Pockets. This procedure is followed when it is desirable to add clearance to the crank end or where additional clearance over and above that built into the head-end head fixed pocket(s) is required. It usually consists of a cast spherical chamber, flanged to fit in place of the usual valve cap, that can be opened or closed by means of a handwheel in the same way as the head-end fixed pocket is operated.

Suction-Valve Unloaders. This method affects cylinder capacity by deactivating or lifting a suction valve that prevents that particular end of the cylinder from compressing gas. For large cylinders, the cylinder capacity and developed horsepower are reduced by approximately one-half. See figure 8.15 for an illustration of this type of equipment.

The amount and type of cylinder clearance is dependent upon the manufacturer and the specific requirements of each job. In general, it is desirable to specify liner-type cylinders so as to provide flexibility in future operations and to allow for future capacity increase to absorb the horsepower that may be obtained through turbocharging an existing engine. In order to provide for flexibility in initial operations, cylinder-liner sizes and clearance arrangements are purchased in such a manner that the cylinder capacity and developed horsepower can be varied either up or down from the design point. The wisdom of providing flexibility in the manner described above has been

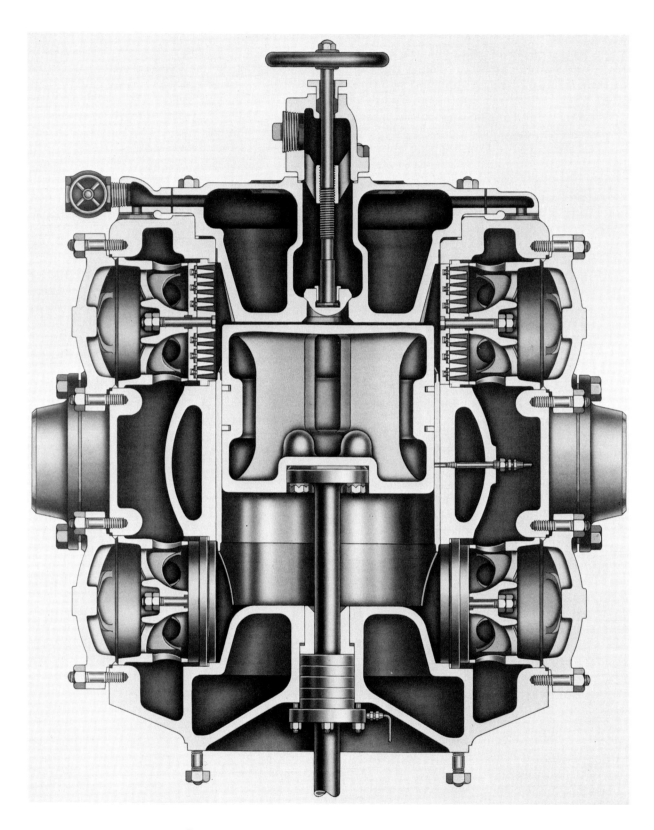

Figure 8.12. Head-End Fixed-Volume Clearance Pocket

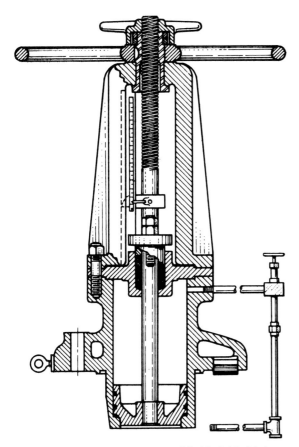

Figure 8.13. Clearance Pocket with Variable Volume

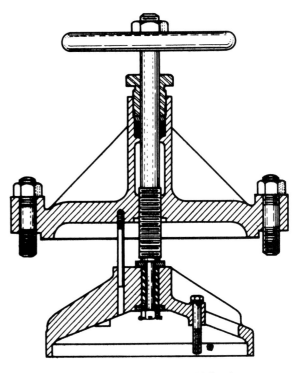

Figure 8.15. Suction-Valve Unloader

proved on many occasions because it is difficult to predict accurately design volumes and pressures prior to plant start-up.

Compressor Selection

There are a variety of factors that influence the selection of one compressor unit over another. Among these factors are the following.

Comparative Costs. Initial purchase price is certainly one of the more important considerations, and with other factors being equal, usually the most competitive offering in this respect is chosen.

Specific Horsepower Required versus Size Available. For some conditions, specific horsepower requirements are such that one manufacturer may make a unit very nearly sized to fit the need and therefore may have a competitive advantage.

Matching Existing Units. Matching a new unit to other equipment already existing in an established plant has many advantages in that operating and maintenance personnel are already familiar with the operation of the proposed new unit and parts interchangeability between units is good which makes the spare part lists not as extensive.

Unbalanced Loads. Manufacturers should supply information concerning unbalanced forces on the

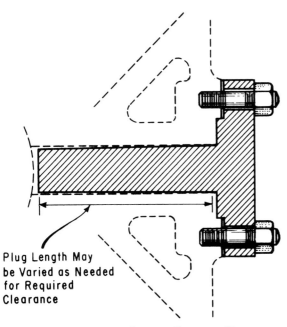

Plug Length May be Varied as Needed for Required Clearance

Figure 8.14. Side-Passage Clearance Plug

gas-engine driver; these include primary and secondary forces and couples both horizontally and vertically. A great amount of unbalance will require more foundation in the best soil areas and will eliminate completely such units for marshy or water sites where the compressor unit may have to be installed on an elevated platform or a pile-supported slab.

Compressor Cylinder Selection and Arrangement. In certain instances, manufacturers will have available units of the same general size and horsepower arrangement, but they will be different as to the number of compressor cylinder spots available, or they will be different as to the cylinder arrangement or amount of horsepower that can be absorbed in one cylinder. Although these situations do not affect unit selection very frequently, they have on occasion been a factor among others in the selection of a specific unit.

Portability. The relative portability between units is sometimes considered. For example, some types of units require two skids and others only one. Naturally, the most compact and portable unit, if the unit can still be maintained easily, will be looked upon more favorably.

Maintenance. Previous experience with different types of compressors may indicate one type requires more maintenance attention than other (e.g., low-speed versus high-speed units). This might have a bearing on the selection of equipment for certain services.

Life of Project. The length of time a compressor might be used at a specific location has a bearing on selection as between portable units and those permanently installed. While either type might be justified in long-lived projects, the portable type is best suited for short-term operations.

Safety Considerations

In accordance with good engineering practice and as insurance against corrosion and other intangibles, all piping within the station boundaries should be designed in accordance with the USAS *Code of Pressure Piping* section 3, "Refinery and Oil Transportation Piping Systems," or section 8 of the same code, "Gas Transmission and Distribution Piping Systems."

Special consideration should be given to the manner in which vent lines are installed so that excess pressure through improper manifolding cannot be applied inadvertently on low-pressure equipment such as starters. Also vents should be installed so that vented gas is carried well away from the unit.

Specific detailed starting and stopping instructions should be posted adjacent to the units, and operators should be required to follow such instructions.

In the design of any installation to operate largely unattended, it is necessary to incorporate certain features of automatic control to protect equipment in case of malfunction. For protection, all engines are equipped with standard shutdown devices for low lube oil pressure, overspeed and high jacket water temperature. The engine can be shut down either by grounding the magneto or by blocking and venting the fuel. The latter arrangement has the advantage that fuel gas cannot accumulate in the exhaust and muffler where it might subsequently detonate or cause a backfire. Additional features may be incorporated at the discretion of the designer to conform with company standard practice or local safety regulations.

A scrubber should be installed immediately upstream of each stage of compression and should include an automatic drain and a device for shutting down the compressor in the event of a high liquid level in the scrubber.

Since a compressor cylinder is designed to handle a certain volume of gas under specific conditions of speed and suction pressure, it is essential to control this pressure within certain limits. For example, if the suction pressure is too high, it is possible to overload the compressor frame or driver as a result of the compressor cylinder handling more than the design volume of gas; if the suction pressure is too low, the discharge temperature of the gas from the compressor might be above safe limits because of the heat of compression and it could thereby result in valve packing, rod or piston failure. Consequently, it may be desirable to install a high-discharge temperature shutdown, particularly if the compressor cylinder is equipped with only one discharge valve on each end. For additional safety, a high-discharge and low-suction pressure shutdown may be provided to function in the event of a gathering line rupture or a station discharge line pressure increase above design limits.

One common practice in connection with control of suction pressure is to automatically vary the speed of the driver to maintain a reasonably constant pressure. With or without speed control, a cycling regulator may be used to divert gas from the discharge side of a compressor and throttle it back into the suction side, thus maintaining a minimum suction pressure to smooth out the engine operation and to assure gas circulation. This procedure has the disadvantage of wasting horsepower, but in many instances it is required to maintain satisfactory compressor operation.

Another practice for preventing excessively high suction pressure is to employ either a pressure-reducing inlet regulator or a back-pressure regulator to flare. In most casing-head gas-gathering systems, either method is satisfactory, but in compressing gas-well gas, the inlet regulator should be used in cases where the gas may not be flared. On installations of the latter type, the wells should shut in automatically and/or the gathering system should be designed for the maximum wellhead pressure.

As a supplement to the usual automatic controls, it is considered wise to have one or more manually operated shutdown stations strategically located safely away from the station. In case of an emergency, these controls can be used to shut down the compressor from a safe distance.

TURBINE–DRIVEN CENTRIFUGAL COMPRESSORS

Gas-Turbine Engines

A gas-turbine engine provides a constant flow of power, contrasted to pulses supplied by a reciprocating engine, by passing hot gases at high velocity across the blades of a turbine wheel. The shaft to which the turbine wheel is attached delivers the power in the form of rotational energy. The basic components of an internal-combustion gas-turbine engine are shown in figures 8.16, 8.17, and 8.18 and are listed below.

1. The compressor. The axial or centrifugal flow compressor provides air under pressure for combustion in the combustor section and, as necessary, for cooling the engine.
2. The combustor. Air from the compressor and fuel at a pressure of about 150 to 200 psi from a fuel pump or gas supply line are brought together and mixed in this part of the engine, and combustion takes place. The hot gases flow to the gas-producer turbine wheel, which is attached to the same shaft as the compressor.
3. The gas-producer turbine wheel. The hot gases passing through the blades of the gas-producer turbine wheel cause it to rotate, driving the compressor and other accessories such as the fuel pump, oil pump, and so forth.
4. The power turbine. The hot gases then pass

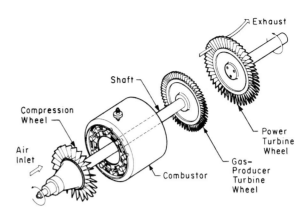

Figure 8.16. Main Components of a Gas-Turbine Engine

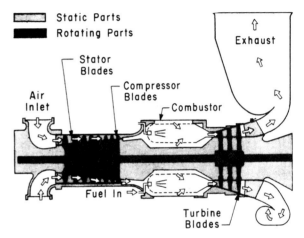

Figure 8.17. Air and Gas Flow in a Gas-Turbine Engine

over another set of turbine blades attached to a shaft. This may be an extension of the shaft to which the gas-producer turbine is attached, or it may be entirely independent. This is the power turbine that provides the power to operate the machine that the engine is designed to drive.

5. Exhaust gases. The hot gases then pass out the exhaust. The remaining heat energy is often used in boilers, oil heaters, and so forth to recover heat that would otherwise be wasted.

Large amounts of air are required by turbine engines in order to control the temperatures in the combustor and turbine sections. About four times as much air is used as would be required for stoichiometric combustion—complete combustion with no excess oxygen. By this means, firing temperatures are held to a range of 1,450 F to 2,200 F. Most industrial

95

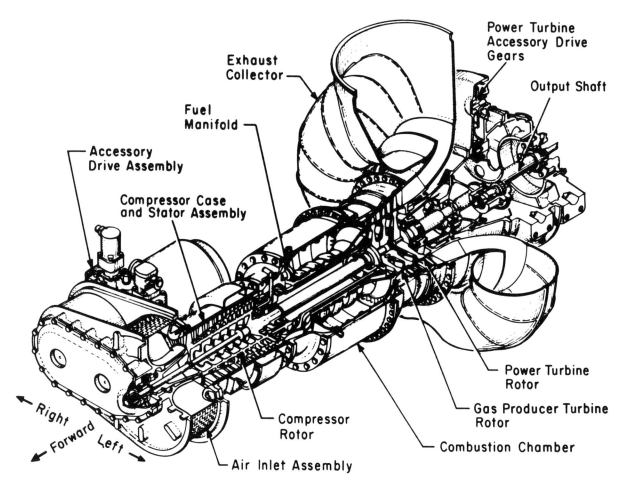

Figure 8.18. Cutaway Drawing of a Gas-Turbine Engine

turbines operate in the lower part of this range, while aircraft turbines operate with the higher temperatures. Exhaust-gas temperatures may run as low as 850 F.

Gas-turbine engines are designed for a wide range of speeds—14,000 to 23,000 rpm being common. They can also operate efficiently at other than design speeds, within reasonable limits. Gas-turbine engines can be directly connected to a high-speed compressor, or the power output can be reduced to a lower speed through gearing to drive a pump or for other low-speed applications.

Gas-turbine engines can operate on kerosene or diesel oil, natural gas, butane or propane, or other fuels. Dual systems are sometimes used for distillate and natural gas. These units adjust easily to changes in the heating value of fuels since the power output is a direct function of the Btu's released in the combustor. Thus, a reduction in heat value of the fuel

will simply result in the fuel throttle opening up to admit a greater amount of fuel. Fuel consumption usually approximates 10,000–13,000 Btu/hp/hr for optimum operating conditions.

In a reciprocating-type engine, power is obtained by burning fuel and the resultant increase in the volume of gases causes an increase in pressure in the power cylinder. This pressure, or force, pushes the piston, and force (F) moves through a distance (D) and results in work (W). Remember, $W = F \times D$. In the case of the turbine engine, fuel is burned in the combustor, resulting in a large increase in the volume of gases; as these gases flow through the turbine wheels to the exhaust, they impinge on the turbine-wheel blades, exerting a force against them and causing the wheels to rotate like fans in a strong wind. Again, a force moves through a distance and results in work. It is not the purpose here to provide a detailed discussion of the theory of gas-turbine

96

engines; such information can be obtained from manufacturers of these units or from standard engineering texts.

The starting sequence for gas-turbine engines is automatic once the operator actuates the starter button. The engine is rotated by an outside power source such as an electric, hydraulic, air, or gas expansion motor for a short period of time to purge the turbine and exhaust system and to place all lubricating systems in operation. Rotation at this point is relatively slow. Next fuel is supplied to the combustor and is ignited. The unit then comes up to speed, and the load is applied. Only a minute or so is required for this entire operation.

Elaborate controls and protective devices are provided to protect against high exhaust-gas temperature, low-lube pressure, high-lube temperature, low-lube oil level, turbine overspeed, and so forth. Control systems are available to permit the completely automatic operation of unattended units in remote locations.

Maintenance requirements on the main part of the gas-turbine engine are relatively minor if the air and fuel entering the unit are free of dirt, water, and so forth. It can be seen readily that such impurities would damage the turbine blades and otherwise adversely affect the close nonwear tolerances necessary for efficient operation. Adequate filtering is absolutely essential. If this is done, major reconditioning of the turbine internal parts needs to be carried out only after intervals of 25,000 to 40,000 hr of operation. Maintenance on the auxilliary equipment and controls needs to be performed at much more frequent intervals, particularly for remote automatically operated units.

Although shaft speeds are high, bearing loads are low due to the steady power flow, lack of load reversals, and so forth. For these reasons, lubricating problems are not great and oils generally have a long service life. However, care should be taken to keep the oil clean and within specification limits. To accomplish this, frequent testing of the lube oil should be part of the maintenance program. The unit manufacturer will provide the lube-oil specifications.

Centrifugal Compressors

One of the laws of physics states that force equals mass times acceleration. Force is ordinarily expressed in terms of pounds. However, pressure, or pounds per square inch, is also a measure of force. This principle is used in a centrifugal gas compressor. Power from the prime mover is used to rotate an impeller, which holds a set of blades. The rotating impeller, by its speed, causes the gas to accelerate and thus to obtain velocity. Since gas has mass, the action of causing it to accelerate results in a force, and pressure is developed in the piping downstream from the compressor.

A centrifugal compressor consists of an outer steel housing containing the internal stationary parts—diaphragms and guide vanes—and a series of stainless steel impellers mounted on a steel shaft located within the housing with the impellers properly spaced with respect to the guide vanes. (See fig. 8.19.) Figures 8.20, 8.21, and 8.22 depict the assembly of these components. Centrifugal compressors have a relatively low pressure ratio, and this requires the use of several stages of compression—four or six stages are common. The great advantage of the centrifugal compressor is its simple construction; it has no valves, gears, or pistons, and the loads are steady and nonreversing. The rotor assembly is the only moving part. However, each multistage unit is limited to a compression ratio of about 3.5 to 1 for natural gas but higher for denser gases such as air. If compression over a greater ratio than this is desired, compressors must be operated in series with intercooling. Maximum working pressure is approximately 1,500 psi. Speeds up to 25,000 rpm are permissible.

Figure 8.19. Impellers for a Centrifugal Compressor

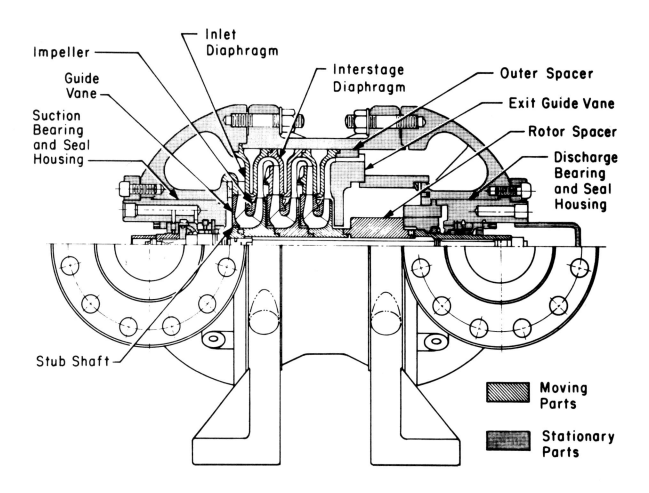

Figure 8.20. Cross Section of a Centrifugal Compressor

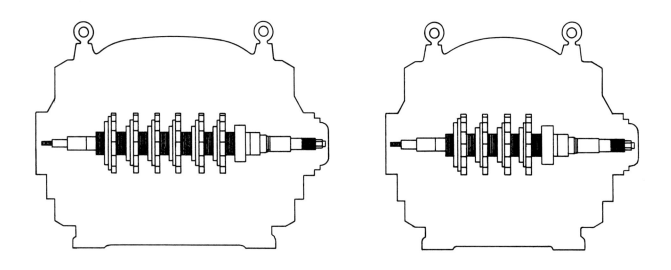

Figure 8.21. Typical Rotor Assemblies for Centrifugal Compressors

98

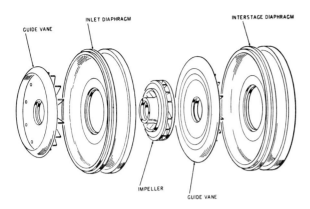

Figure 8.22. Internal Parts of a Centrifugal Compressor

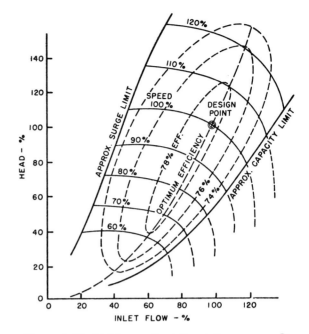

Figure 8.23. Multistage Centrifugal Compressor Performance Curves

Figure 8.23 illustrates the performance of typical multistage compressor. It will be noted that as flow through the compressor is decreased from design capacity but at a constant pressure ratio, a point is reached where stable operation is no longer possible. This is commonly called the surge limit and is characteristic of all centrifugal compressors. Operation in this range is to be avoided. In designing units, staging is selected to allow an acceptable margin between the design point and surge limit.

Centrifugal compressors must be protected from impurities such as dirt and liquids. Dirt will cause premature wear, and liquids can seriously damage the internals such as impeller, guide vanes, and so forth.

Figure 8.24 shows an installation of two 1,000-hp turbine-driven compressors in service at a South Texas location to boost pipeline pressure from 500 to 1,000 psi.

Figure 8.24. 1,000 HP Turbine-driven Centrifugal Compressors

99

IX

INSTRUMENTS AND CONTROLS

All functions described in this chapter are related to the control of a gas or liquid; these two items are referred to as the controlled medium. The end result is to maintain the controlled medium at a specific pressure, temperature, flow rate, or liquid level. These last items then are referred to as controlled variables. Final control is accomplished either directly or indirectly by increasing or decreasing the size of the opening through which the controlled medium or some allied agent is passing. Most protective devices for gas compressors and other field equipment operate on these same principles.

MEASUREMENT OF CONTROLLED VARIABLES

Before any variable can be controlled, its value must be determined and compared to an optimum value, in order to take any needed corrective action. Corrective action may be made manually by operating personnel after visually checking the value of the variable as indicated by a measuring instrument or an electrical or mechanical device. The operating personnel compare the measured value to an optimum value and automatically make the required correction.

Pressure Measurements

Although pressure can be any force from less than atmospheric (vacuum) to thousands of pounds above atmospheric, the discussion will be limited to a measurement above atmospheric. However, the instruments described can also be used to measure vacuum.

The *manometer* is the most commonly used instrument for measuring pressures from zero through a few pounds per square inch. This device is a U-shaped tube made of glass or other transparent material as illustrated in figure 9.1. The inside diameter of the tube is unimportant provided the size is consistent over its entire length. A scale with suitable markings is attached behind the unit, and the tube is filled approximately half full of fluid. Each scale is marked for a fluid having a particular specific gravity, and a fluid of any other gravity will give an incorrect reading. One end of the manometer is connected to the pressure to be measured, and the other end is open to atmosphere. The pressure will force the fluid to move in the tube until the weight of the displaced fluid is equal to the force exerted by the pressure. The difference between the height of the fluid in the two sides of the manometer is the pressure measurement. Scales are generally marked in millimeters,

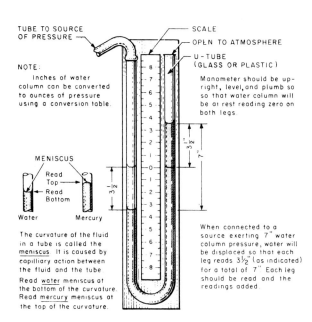

Figure 9.1. Typical Water Manometer

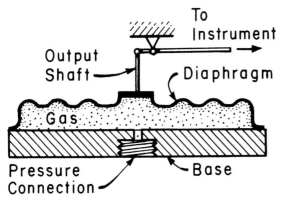

Figure 9.2. Single-Diaphragm Pressure-sensing Element

inches, or ounces. Manometers are used almost exclusively in measuring pressure where visual readings are required and low pressure is involved.

A *diaphragm element* is also used in measuring low pressures up to a few pounds. It is extremely sensitive to small pressure changes and can provide an indicated reading or the motion required to operate an automatic control device to be discussed later. The basic diaphragm unit is shown in figure 9.2 and consists of a heavy metal base, a pressure connection, and a corrugated diaphragm. The diaphragm is generally made of thin sheet metal formed to shape by a

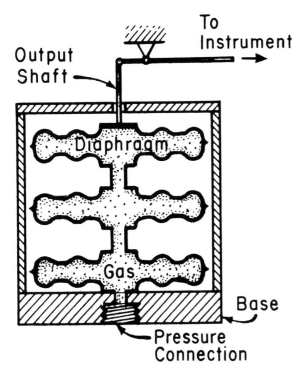

Figure 9.3. Multiple-Diaphragm Pressure-sensing Device

hydraulic press. However, nonmetallic materials are available to measure extremely low pressures. Corrugations allow considerably more movement than could be expected from a flat sheet for any applied pressure. Pressure applied inside the unit expands the diaphragm, and this motion is transmitted through suitable linkage to an indicating pointer, a recording pen, or the detection mechanism of an automatic controller. A special type of diaphragm element is used for applications such as aneroid barometers. If the pressure inlet is sealed, any variation in atmospheric pressure acting on the outside of the diaphragm will cause a pointer movement proportional to the change in atmospheric pressure. The motion from a single-diaphragm unit is limited, and the process of magnifying this motion with additional linkage can introduce errors and create problems of calibration. The additional movement needed to position a pointer along a longer indicating scale can be achieved by adding additional diaphragms in series to obtain an extremely sensitive and accurate measuring device (fig. 9.3).

A *bellows* unit is similar to the diaphragm element except that the corrugations extend in series rather than expanding outward. In figure 9.4, the movement of the bellows resulting from an increase in pressure is opposed by a spring. This spring extends the range of pressure a particular size bellows is capable of measuring. Again, the movement of the bellows positions an indicating pointer, a recording pen, or the detection mechanism of an automatic controller. These units are available with ranges up to 100 lb.

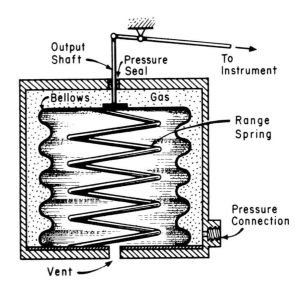

Figure 9.4. Bellows Pressure-sensing Element

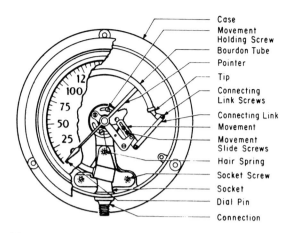

Case
Movement Holding Screw
Bourdon Tube
Pointer
Tip
Connecting Link Screws
Connecting Link
Movement
Movement Slide Screws
Hair Spring
Socket Screw
Socket
Dial Pin
Connection

Figure 9.5. Bourdon Tube Pressure-indicating Instrument

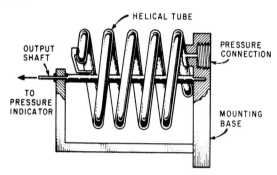

HELICAL TUBE

OUTPUT SHAFT

PRESSURE CONNECTION

TO PRESSURE INDICATOR

MOUNTING BASE

Figure 9.6. Helical Tube Pressure-sensing Element

A *Bourdon tube* is essentially a C-shaped tube that is sealed at one end and equipped with a pressure connection on the other end. As shown in figure 9.5, pressure introduced into the tube will apply an equal force on the entire inner surface of the tube. Since the outer circumference is greater than the inner circumference, pressure will tend to straighten the tube. The movement of the sealed end of the tube is used to position a pointer, a recording pen, or an error-detecting unit. The amount of movement of the sealed end is directly proportional to the portion of a circle formed by the tube. A normal Bourdon tube will have a small movement for minor changes of pressure, and this motion must be amplified with suitable linkage if a long scale is needed. This again increases calibration problems and the possibility of error. If the tube is wound in a spiral, more tip movement will result from a pressure change. If the spirals are wound in a helical shape as shown in figure 9.6, each spiral is of the same circumference and the accuracy of the tube is improved considerably. Each type of tube is available for measuring pressures up to thousands of pounds.

Temperature Measurements

The common indicating *thermometer* is the most widely used instrument for measuring temperature. It consists of a bulb filled with liquid—which may be mercury, pentane, toluene, alcohol, and the like—connected to a glass stem. Changing temperature causes the liquid to expand or contract and move up or down the glass stem thus giving a visual indication of temperature. The bore of the stem is very small, and the volume of fluid in the bulb is about 1,000 times the volume of the stem bore.

In order to check the temperature of a flowing stream of gas or liquid, a method must be provided to get the thermometer bulb inside the pipe. The *thermometer well*, illustrated in figure 9.7, is permanently mounted in the piping system. When the temperature is checked, the thermometer is inserted in the well and packing is installed in the annular space around the stem in order to eliminate variations due to ambient temperature. The well should be filled with a suitable liquid, such as mercury, in order to get good heat transfer between the material and the thermometer.

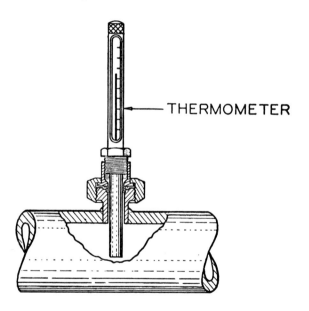

THERMOMETER

Figure 9.7. Thermometer Well

A filled-system thermometer, as shown in figure 9.8, is used to provide a mechanical movement needed to position an indicating pointer, a recording pen, or an error-detection unit in an automatic controller. The system consists of a bulb, small diameter connecting tubing, and a hollow spiral tube. The entire system can be filled with a fluid such as

102

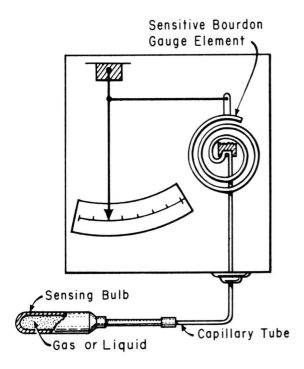

Sensitive Bourdon
Gauge Element

Sensing Bulb

Gas or Liquid

Capillary Tube

Figure 9.8. Filled-System Thermometer

mercury or a gas. Other systems are partially filled with a fluid that tends to vaporize at the temperature range being measured.

Systems completely filled with a fluid or gas depend directly on the expansion and contraction of the fill medium to position the tip of the spiral tube. Since ambient temperature affects the fill medium in the connecting tubing and the spiral elements, this type of system must have a compensating device.

Compensation is usually accomplished by a second spiral element connected to a dead-end tube (fig. 9.9). The two elements are spiraled in opposite directions in such a fashion that the free end of the compensating element provides the mounting location for the measuring element. Thus, any movement in the measuring element occurring as a result of an ambient temperature change will be exactly corrected by the compensating spiral.

Systems partially filled with a fluid depend on the change in vapor pressure resulting from temperature changes to operate the spiral tube. In this type of unit, the tube constitutes a pressure element measuring vapor pressure and it is not effected by changes in ambient temperature. However, the measuring range for each type of fluid (ethyl alcohol, ether, propane, butane, etc.) is rather limited, and the boiling point of the fill fluid must be lower than the lowest temperature to be measured in each case.

A *bimetal temperature element* is comprised of two different metal strips. Every metal or alloy will expand or contract with temperature changes; however, the rate of expansion will vary from one metal to another. If two dissimilar metal strips are bonded together along their length and held rigid at one end, the free end of the unit will move as the temperature changes. As the temperature increases, one strip of metal will expand to a greater length than the other. However, since the two strips are bonded together, the entire unit must bow in such a fashion that the longer strip occupies the outer circumference of the arc. A reverse action will occur as the temperature decreases. (See fig. 9.10.) The movement of the free

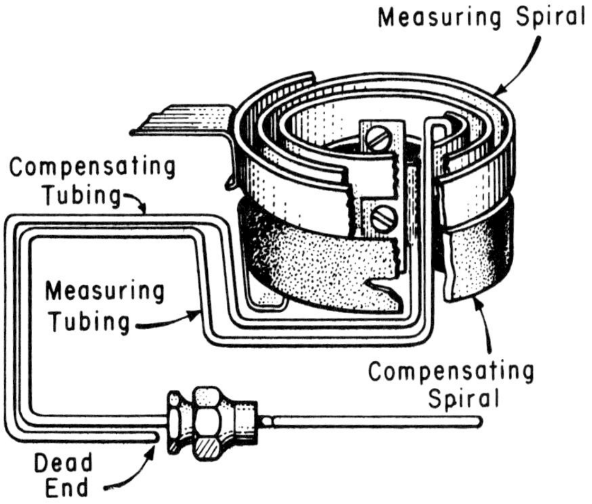

Measuring Spiral

Compensating
Tubing

Measuring
Tubing

Compensating
Spiral

Dead
End

Figure 9.9. Compensating Filled-System Thermometer Element

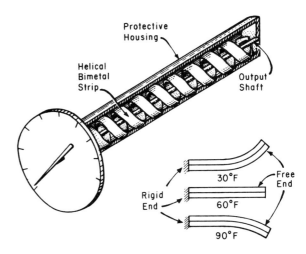

Protective
Housing

Helical
Bimetal
Strip

Output
Shaft

Rigid
End

Free
End

30°F

60°F

90°F

Figure 9.10. Bimetal Element for Temperature Measurements

103

end of the unit is directly proportional to the length of the bimetal strip. To conserve space, most temperature-sensing strips are normally wound in a spiral or helical shape. These units are generally used as indicating devices or for safety units.

A *thermocouple element* consists of two different metals attached together. When any type of metal is heated, negative particles of electricity (electrons) are driven away from the heat. The number of electrical charges moved will vary in different types of metal. If two dissimilar metals are joined at one end and the junction heated, electrical charges will move down each metal and accumulate at the free ends. However, there will be a greater accumulation at one free end than at the other due to differences in the two metals; the difference in accumulated charges is directly proportional to the heat applied to the junction between the two metals. (See fig. 9.11.) The difference in the electrical charge at the free end amounts to only a few thousandths of a volt, even at extremely high temperatures of 3,000 F or more. If a very sensitive voltmeter is connected to a free end, it can measure this small voltage. Even though the meter measures voltage, the scale is marked to read temperature. Thermocouples can indicate temperature directly, or the electric output can be used to energize an electric temperature controller.

Liquid-Level Measurements

The most commonly used device for a direct reading of liquid level is the sight glass. This unit consists simply of a transparent column, usually glass, suitably attached to the tank containing the liquid. When the valve between the bottom of the column and the tank is opened, fluid will rise in the column to the same height as the fluid in the tank. This level can then be compared to the scale attached to the column. If the vessel is at atmospheric pressure, the top end of the column can be vented. However, on pressure vessels, both ends of the column must be connected to the tank or fluid will be blown out the open end. Check valves must be used in conjunction with pressurized vessels to prevent fluid leakage if the glass is accidentally broken.

A *float-type unit* can be used when liquid-level control is needed. One such unit employs a bulb or ball that floats on the surface of the liquid. As the liquid level varies, the float element changes position. This motion is transmitted through suitable linkage to a shaft and through a packing gland to the outside of the vessel where it is used to position an indicator or liquid-level controller. The float assembly can be

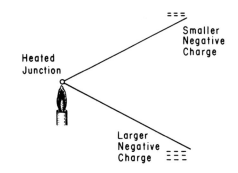

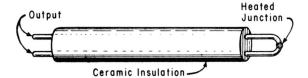

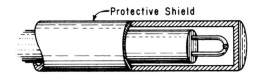

Figure 9.11. Thermocouple Temperature Element

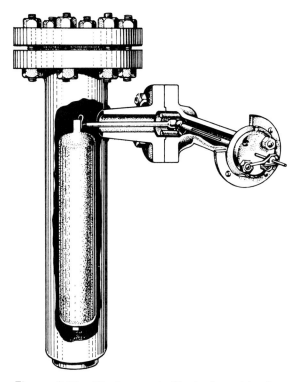

Figure 9.12. Displacement Float Assembly for a Liquid-Level Controller

104

flanged directly into the vessel or into a float chamber, which in turn is connected to the vessel. The range of levels that can be measured with this device is limited since the float rides on the surface of the liquid. It is generally used for on-off dump-valve applications. Another assembly is designed in such a fashion that the liquid displaced weighs less than the bulb itself. Consequently, the float sinks in the liquid as the level rises. However, the weight of the float is reduced by an amount equal to the fluid displaced.

Figure 9.12 shows the float attached to the float-rod assembly. The float rod is held in position by a support bracket. A hollow torque tube with a sealing flange welded on one end and a suitable socket arrangement welded on the free end fits over the float-rod assembly. A small steel shaft welded to the socket assembly on the free end of the torque tube extends down the center of the tube and through the sealing flange to the outside.

When the vessel is empty, the full weight of the float causes the torque tube to twist and rotate the shaft leading outside. As the fluid fills around the float, the weight available to twist the torque tube is reduced by an amount equal to the fluid displaced by the float. The tube will move toward its normal position and rotate the shaft. This shaft rotation is used to supply information to the sensing device of a controller.

The force exerted by a liquid on the containing vessel is directly proportional to the depth of the fluid at the measuring point. A bellows or diaphragm mechanism can be used to measure this pressure. Figure 9.13 is a view of a diaphragm used to measure

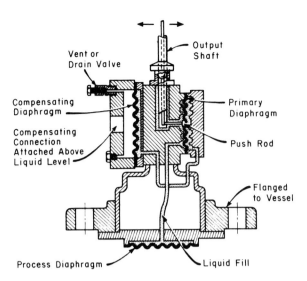

Figure 9.14. Diaphragm Mechanism for a Pressure-compensating Liquid-Level Controller

this force in a vessel vented to atmosphere. This arrangement is unsatisfactory for a pressurized vessel since the diaphragm would record not only the liquid pressure due to depth but any additional gas pressure inside the vessel as well. Figure 9.14 shows the arrangement used on pressurized vessels. The two bellows or diaphragms oppose each other. Both the fluid-level pressure and the gas pressure are applied to the bellows connected to the bottom of the vessel while only gas pressure is applied to the bellows connected to the top. Thus the entire arrangement measures pressure difference between the top and bottom of the vessel, which is the force produced by liquid level. The bellows movement of either arrangement is used by the error-detecting unit of a controller.

FLOW CONTROLLERS

Manual Control

Most control functions can be accomplished manually by a well-informed operator. However, all control functions can be accomplished better, safer, more accurately, and at less cost by the proper type of automatic control. To maintain control of a process, each automatic system must contain the following essential elements:

1. a measuring element to detect any deviation in the controlled variable;

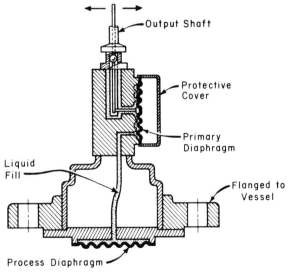

Figure 9.13. Diaphragm Mechanism for a Liquid-Level Controller

2. a controlling element that utilizes any detected error to produce a control signal; and

3. a final control valve that utilizes the control signal to position the valve at the opening required to maintain the controlled variable within acceptable limits.

The process and the automatic-control instruments are connected in a continuous closed loop giving the process the ability to control itself. Setting the acceptable limits on the controlled variable is the only manual operation required. The measuring and controlling elements are usually both incorporated in a single unit. The final control valve is usually positioned by an electric or diaphragm motor.

Flow Regulators

Every regulator, regardless of size or construction, contains three essential elements. They are (1) a restricting device to limit the flow of product through the unit, (2) a measuring device to determine the value of the controlled variable, and (3) a loading device to reposition the restricting device when necessary. A self-contained regulator contains all of these essential elements within one unit.

The *spring-loaded regulator* is most commonly used to supply fuel gas to line heaters and other burners found in the field. They are also used to supply low-pressure gas for household use. Figure 9.15 is a view of a spring-loaded regulator in its operating position. The loading element in this case is a compressed spring. This spring is applying a force to the top side of a flexible diaphragm (measuring element) causing the diaphragm to move down. This motion moves the toggle assembly and allows the seat pad to move away from the orifice.

When high pressure is admitted to the inlet of the regulator body, the gas will pass through the open

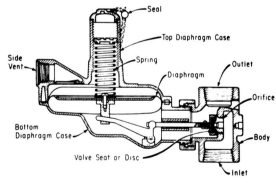

Figure 9.15. Spring-loaded Regulator

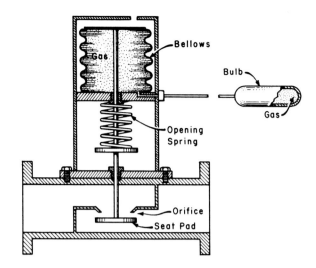

Figure 9.16. Bellows-Type Temperature Controller Device

orifice and into the downstream piping causing higher pressure. This pressure increase is felt by the diaphragm since the chamber below the diaphragm is open internally to the outlet piping system. The diaphragm will move up when the force exerted by the gas pressure exceeds the spring tension. This movement closes the seat pad against the orifice to restrict or stop the flow of gas. Changing the spring tension will change the outlet-pressure setting.

Regulators are available in this category that will accept inlet pressure above 1,000 lb and provide outlet pressure up to about 400 lb. The unit has the disadvantage of having the outlet pressure applied to the diaphragm and casing, and this is the limiting factor on maximum outlet pressure. In addition, the diaphragm must move down to open the orifice. This allows the spring to extend somewhat and relieves part of the tension. As a result, the outlet pressure drops off as the regulator opens farther.

A unit designed to control temperature is shown in figure 9.16. The diaphragm has been replaced by a bellows unit connected to a bulb. As the temperature increases, the fluid inside the bulb expands. This expansion is transmitted to the bellows repositioning the seat pad. It should be noted that this unit only controls the flow of the heating agent (gas, steam, or hot water) and totally disregards pressure. The pressure must be regulated through a separate unit.

Pilot-operated regulators operate on the same basic principle as the spring-loaded unit except that the loading force is a constant gas pressure rather than a spring. This constant pressure is applied to the top side of the diaphragm by a small spring-loaded pilot regulator. Most pilot-loaded units employ a closing spring to aid in holding the seat pad against the orifice. As a result, the loading pressure must overcome both the outlet pressure beneath the diaphragm and the closing-spring tension in order to open the orifice. Changing the spring tension on the pilot regulator will change the loading pressure and the outlet pressure. This unit has a constant loading force regardless of the position of the diaphragm, and this eliminates the drop in pressure associated with an increase in flow through a spring-loaded unit. However, the full outlet pressure is still applied to the diaphragm case, and this limits the maximum outlet available. These units find their greatest use in supplying gas to distribution systems and in large burner applications requiring a more stable pressure than can be delivered by a spring-loaded assembly.

Motor Valves

A diaphragm motor valve is the most common unit used for final control. Referring to figure 9.17, a

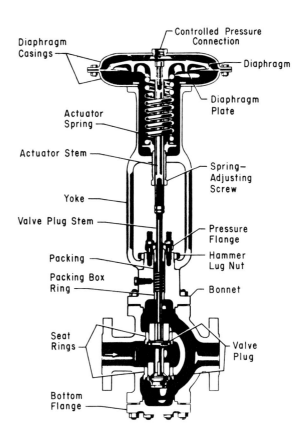

Figure 9.18. Normally Closed Diaphragm Motor Valve

valve of this type consists essentially of a diaphragm, loading spring, valve stem, valve body, inner valve, and valve seat.

The loading spring attempts to hold the inner valve either completely open or completely closed. The first is called a normally open valve (fig. 9.17) and the latter is termed a normally closed valve (fig. 9.18). Operating pressure (control signal) is applied to the top side of the diaphragm overcoming the resistance of the loading spring and increasing or decreasing the valve opening, depending on the arrangement of the valve seats. The pressure required to start compressing the spring is generally 3 lb; the spring will then require 15 lb of pressure to achieve maximum compression. On some valves, this pressure range will be between 6 and 30 lb. In case of complete loss of control signal, the loading spring immediately returns the valve to its normal position of open or closed.

A different arrangement for a motor valve is shown in figure 9.19. In this device, a diaphragm is still used but is attached to a hydraulic piston rather than a metal stem. An increase in the control signal moves the diaphragm down, building pressure in the

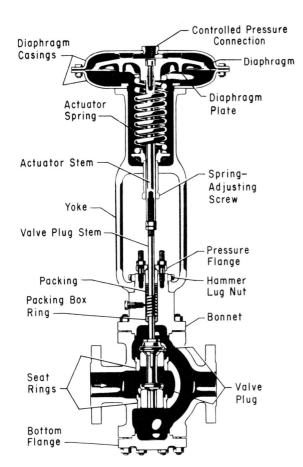

Figure 9.17. Normally Open Diaphragm Motor Valve

107

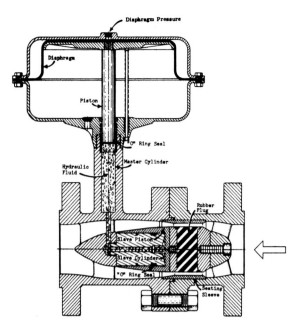

Figure 9.19. Hydraulic Piston Motor Valve

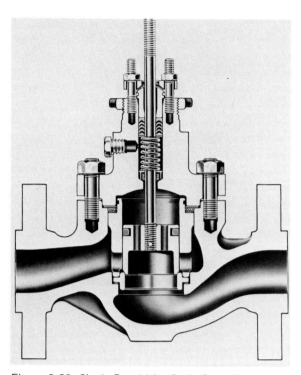

Figure 9.20. Single-Port Valve Body for a Diaphragm-actuated Motor Valve

hydraulic system. This hydraulic pressure is applied to a slave piston inside the body, moving the slave forward against a rubber plug. Since the plug is mounted solid at the front end, the forward movement of the slave piston deforms the rubber so that it extends to the side and contacts a smooth seating sleeve inside the regulator body. This in turn stops the flow. The flow through this body is streamlined since there is little change in direction of the flowing material. In addition, the rubber plug is capable of sealing off around trash that would normally hold other types of inner valves open. The plug will withstand a pressure drop of over 1,000 lb and still go to a complete shutoff. Unlike the conventional motor valve, the control signal range required to operate the valve will depend on the pressure drop across the plug, but in most installations, it will be less than the 3 to 15 lb mentioned previously.

In addition to being either normally open or closed, the inner valve of a conventional motor valve can be either single or double ported. Figure 9.20 is a single-ported valve as opposed to the double-ported valves in figures 9.17 and 9.18. The double-ported unit is sometimes referred to as a balanced valve since the material passing through the body tends to force the inner valve open at one port with the same force as it applies to the other port in a closing direction. Since the inner valve is balanced regardless of its relative position to the seat rings, the double-ported unit is used primarily when the process requires

extended throttling action at less than 10 percent of capacity. However, it is very difficult to achieve a complete shutoff. The single-ported arrangement is unbalanced and not suited for service requiring a throttling of the inner valve. Since there is only one seating surface in the single-ported unit, it is used primarily when a complete shutoff is required.

Inner valves are produced in various forms, each with a particular flow characteristic in mind. Five of the most widely used valve forms are shown in figure 9.21, and the relationship between valve movement and percent of maximum flow is shown in the chart. The linear plug provides the most linear flow characteristics; that is, 20 percent of valve travel provides 20 percent of maximum capacity, 40 percent travel provides 40 percent of capacity, and so forth. This form provides good, general-purpose control. The V-port form provides excellent throttling characteristics since 40 percent of valve movement will only result in very small changes in flow. This is especially beneficial when the controlled variable can be changed or upset easily. The quick-opening valve provides a large change in flow for small changes in valve position in the throttling area. This form is used when drastic changes are expected in the controlled variable or when the controlled variable is difficult to change.

108

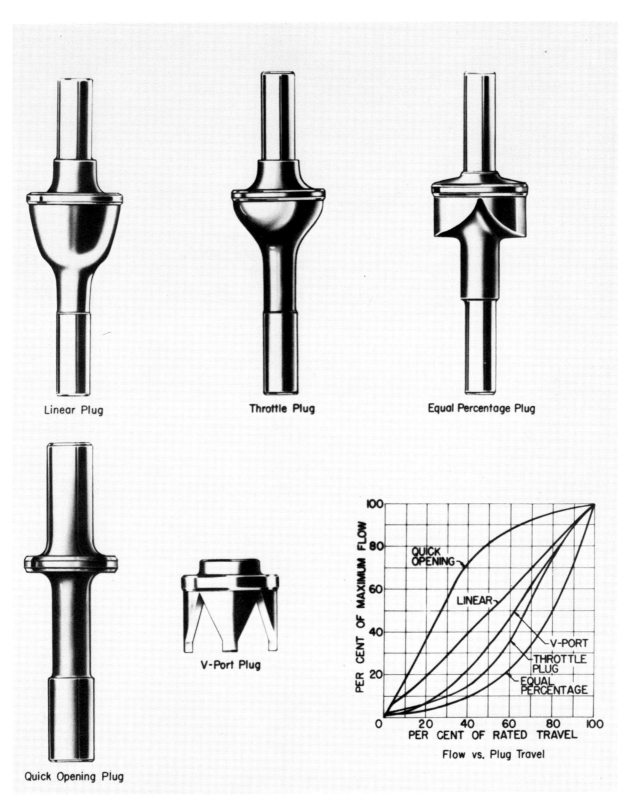

Linear Plug

Throttle Plug

Equal Percentage Plug

Quick Opening Plug

V-Port Plug

Flow vs. Plug Travel

Figure 9.21. Various Valve Forms for Motor-operated Control Valves

109

Motor valves operate with equal effectiveness in either the normally opened or closed configuration. The choice between the two is determined generally by the action that can be tolerated best in the event of control-signal failure. If the installation controls pressure in a transmission line or a distribution system, a normally open valve would be used. Control-signal failure would allow the valve to go open, build the pressure somewhat, and blow a relief valve. However, customer service from the line would not be interrupted. On the other hand, this type of operation could be potentially dangerous in a processing plant. Escaping gas creates a fire hazard, and excessive pressure, temperature, or flow generally results in a loss of product. Therefore, the normally closed valve is used for most purposes in a processing plant.

The choice between a single- or double-ported valve is determined by the job to be performed and the operating conditions. Long transmission lines or distribution systems contain millions of cubic feet of storage space downstream of the valve. The valve must be capable of delivering large volumes of gas, generally is not required to throttle for extended periods of time, and never goes to a complete shutoff. This situation is ideal for a double-ported valve equipped with a linear plug. If the unit is controlling flow instead of pressure, the throttle plug would be more suitable since small changes in valve position would not change the recorded differential so drastically.

If the controlled variable is reduced two or more times in a relatively short section of piping, there will be a very small amount of storage area between the valves. The controlled variable can be changed very easily; therefore, a throttle-plug inner valve must be used if stable conditions are to be maintained. On installations that require a valve to be either wide open or completely closed (liquid level, dump valve, etc.), the single-ported unit equipped with a quick-opening plug is satisfactory. The same unit may be required for applications handling corrosive material that would erode any type of double-ported assembly.

Pneumatic Controllers

Diaphragm motor valves must be equipped with a pneumatic controller in order to automatically control any process. In addition, the controller must be furnished a supply of clean, dry, noncorrosive air or gas to be used as a control signal. In the field, gas is used almost exclusively. The supply must be regulated to a constant pressure (usually 20 lb) before being introduced into the controller.

Controllers provide the following functions related to automatic control:

1. a method of setting the value to which the controlled variable is to be maintained;
2. a device to measure the actual value of the controlled variable;
3. a system to detect any deviation of the controlled variable from the desired setting; and
4. a system to react to the detected deviation and correct the control signal.

In most of the newer or more sensitive controllers, an amplifying device is also employed to boost the small corrective signal produced by the controller to a level that will operate the valve more effectively.

Any device that can determine the value of the controlled variable can be considered a measuring element. This includes Bourdon tubes, bellows, temperature elements, manometers, flow meters, and so forth. In order to get the most sensitive control from the equipment, a measuring element should be chosen with the lowest maximum range that will accommodate the process without having to operate at the extreme end of the range. In other words, never use a 600-lb element to control at 50 lb or a 100-lb element to control at 95 lb. The element must be constructed of a material that will not be adversely affected by the product or atmosphere.

Deviation of error detection is accomplished in most controllers by a nozzle and flapper system. Figure 9.22 is a sketch of a very simple controller using a Bourdon tube to measure the controlled variable. Suitable linkage connects the free end of the Bourdon tube to the flapper at point "A." As the controlled pressure changes, the Bourdon tube expands or contracts and moves the flapper at point A. This in turn causes the flapper to rotate about the pivot point and partially open or close the small orifice at B. The error is determined by the distance between the flapper and nozzle. Once the error has been detected, a corrective signal is generated by two orifices, one in the supply line and one at the nozzle tip. A constant 20 lb of pressure is supplied to the primary restriction by a small pilot regulator. This primary restriction allows only a specific maximum flow into the control system. The hole in the nozzle is larger than the primary restriction; consequently with the flapper away from the nozzle, the flow out of the control system can be greater than the

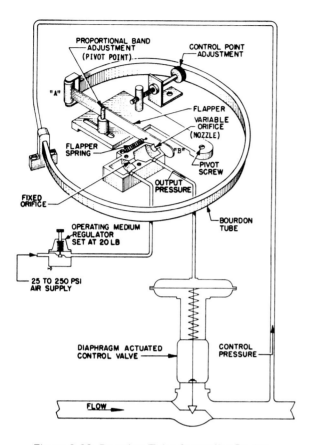

Figure 9.22. Bourdon Tube Controller Device

incoming flow and pressure cannot build. This control system is tied into the diaphragm chamber of the motor valve. If the valve is normally open (i.e., diaphragm pressure is required to close the valve), the inner valve will be completely open. If gas is allowed to flow through the valve, the outlet pressure will increase and cause the Bourdon tube in the controller to expand, rotate the flapper about the pivot point, partially close the nozzle, and restrict the flow out of the control system. This restriction of the flow out of the control system allows pressure to built up on the motor-valve diaphragm and close the inner valve. In normal operation, the nozzle is neither completely open or closed, but instead the nozzle is restricted to a point of keeping the control signal at the level required to hold the valve closed enough to maintain the desired downstream pressure. The value of the controlled variable is adjusted by changing the distance between the flapper and nozzle; in most instances, the adjustment is achieved by moving either the pressure element or the nozzle assembly. If the controller is installed on a normally closed valve,

the control action can be reversed; that is, an increase in the controlled variable will result in a decrease in the control signal. This reverse action will occur if the nozzle is moved to the opposite side of the flapper.

Proportional Action

When a controller is installed, what type of response is required? The answer will depend largely on the system under control; how much variation from the desired value is permissible, how much piping is downstream of the regulator, and so forth. Of course, the ideal situation would be no variation at all, but in most instances this snap-action type of control cannot be tolerated except in liquid-level control. If the controlled variable goes above or below the desired value by a slight amount, the inner valve will go completely open or closed. This results in a very erratic control pattern, and in most applications the reaction of the regulator to a change in controlled variable must be restricted. This is accomplished by adding proportional action to the controller. *Proportional action* simply means the amount of change in the controlled variable required to completely stroke the inner valve from one extreme position to the other. Proportional settings are generally given in percentage, and this denotes the percent of the total control range required to stroke the valve. In a hypothetical case, a controller could have a 100-lb spring range and a proportional setting of 10 percent. A change in the outlet pressure of 10 percent of the 100 lb (or 10 lb) would be required to stroke the valve. If the valve shuts off completely when the controlled pressure reaches 50 lb, the pressure will drop to 40 lb before the valve will completely open. Of course, at 45 lb the valve is halfway open.

In several controllers, changing the proportional setting is accomplished by moving the pivot point along the flapper. In figure 9.22 if the pivot point is moved near point *A*, a small change in outlet pressure will result in a small movement of point *A* but a large movement at the nozzle end of the flapper due to the pivot location. This in turn will completely stroke the valve. The controller is at 0 percent proportional setting. If the pivot point is moved to a position near the nozzle, a great deal of change in the controlled pressure is required to get enough movement at point *A* to bring the flapper against the nozzle. The proportional setting is now 100 percent.

111

Setting Proportional Action

If there could be an instantaneous repositioning of the valve whenever a deviation in the controlled variable occurred, operation at very near 0 percent proportional setting could be achieved. However, some lag will always exist since the deviation must first be detected and a corrective control signal generated before the valve is repositioned. Such things as the size of controller-restricting orifices and the volume within the control system (i.e., the connecting tubing and diaphragm case) affect the maximum speed at which the valve can be repositioned. This maximum repositioning speed will remain constant for any particular installation. The proportional setting is made in accordance with this lag time. Adjustment of proportional action is started at a high setting. The controlled variable will remain stable, but a large deviation from optimum value can exist. As the proportional setting is brought toward 0 percent, the deviation from optimum value will reduce. The adjustments are made in small increments with enough time between adjustments to allow the process to stabilize. Eventually a setting will be reached in which the required change in variable occurs faster than the valve can be repositioned, and the process will be unstable. The proportional setting is then increased enough to stabilize the system for maximum response speed.

A more elaborate control system is shown in figure 9.23. This unit is equipped with a constant supply to a primary restriction, a secondary restriction at the nozzle that is larger than the primary restriction, a measuring element, a flapper, and a pivot point. The arrangement is a little different. A bellows opposed by a spring has been added, and a piece of tubing ties the bellows into the motor-valve control system. Any increase at all in the controlled variable is enough to bring the flapper against the nozzle. As the pressure on the diaphragm builds, the change is felt by the bellows and it extends against the spring forcing the beam down at point B. The beam rotates about the pivot point and lifts the flapper away from the nozzle, opposing the action of the measuring element. The amount of opposition depends on the position of the pivot point. If the pivot is near A, movement of the bellows has little effect on the flapper. Under this condition, a slight change in controlled variable is enough to stroke the valve, and the proportional setting is 0 percent. If the pivot is moved near B, the bellows movement causes a great deal of opposition to the flapper movement created by the measuring element and a large change in variable is required to stroke the valve. This type of controller can be designed to go up to 400 percent proportional adjustment.

Reset Action

The controllers described previously can only have the proportional setting as close to 0 percent as possible without creating an unstable control condition. This means the controlled variable must deviate from the optimum level, depending on load conditions, in order to open the valve. In some applications, this deviation cannot be tolerated. Automatic

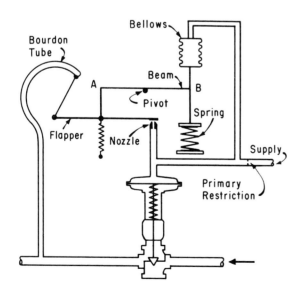

Figure 9.23. Wide-Range Proportional Controller

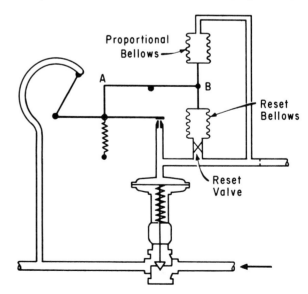

Figure 9.24. Proportional Controller with Reset

reset allows the slightest change in controlled variable to stroke the valve completely, if necessary, and to regain the optimum value without getting into the inherent erratic control associated with normal 0 percent proportional action. In effect, reset action makes the controller operate at 0 percent proportional band regardless of the actual proportional setting. However, the controller reaction to a change in variable is retarded to prevent the erratic control.

Figure 9.24 is a control system very similar to that just described. However, the proportional bellows is opposed by a reset bellows rather than a spring. This reset bellows, along with the proportional bellows, is tied into the control system. A change in the pressure to the diaphragm is felt by the proportional bellows immediately while the change is delayed to the reset bellows by the reset valve. This reset valve is a very fine needle valve. With the pivot point moved under point *A*, the controller operates at 0 percent proportional band. Any change in the controlled variable is enough to cause the controller to stroke the valve since the feedback and movement of the bellows causes no movement of the flapper. The same condition exists if the beam remains in one location or can be brought back to an original position. The following set of circumstances might exist with this type of controller:

1. 100 lb pressure spring;
2. 3- to 15-lb diaphragm-pressure change required to stroke motor valve;
3. 10 percent proportional setting;
4. 1 repeat per minute setting on reset valve; and
5. system stable with 14 lb on entire diaphragm-pressure system including the proportional and reset bellows, making the beam completely level at this time.

If due to additional consumption the controlled variable drops off 1 lb from the ideal value or 1/10 of the amount required to stroke the valve using proportional action alone, the pressure element will cause the flapper to move away from the nozzle and vent diaphragm pressure. This lower diaphragm pressure is felt immediately by the proportional bellows but not by the reset bellows because of the reset valve being almost closed. Consequently, the pressure in the reset bellows remains at the original pressure temporarily, acts as a spring, forces the beam up at point *B* and down at point *A*, and opposes the action of the pressure element. Under the circumstances stated in the example, there would be a net drop in diaphragm pressure initially of 1.2 lb due to proportional action alone. So far, the response is the same as

encountered in a controller with proportional action only. In that case, the proportional bellows was opposed by a spring that applied a constant force. In this application, the entire diaphragm-pressure system, with the exception of the reset bellows, has dropped from 14 lb to 12.8 lb. There is a 1.2-lb difference in pressure between the two bellows due to initial proportional action. Without reset action, the controller would maintain this 12.8 lb on the control system as long as consumption remained at this new rate.

The pressure gradually bleeds out of the reset bellows. As this pressure bleeds out, the force exerted against the proportional bellows decreases and the beam moves down at *B*, moves up at *A*, and pulls the flapper away from the nozzle. This, in turn, reduces the pressure in the rest of the diaphragm system and the same 1.2-lb unbalanced pressure condition exists between the reset and proportional bellows. Proportional action can either increase or decrease this unbalance if the control variable changes again. The pressure in the diaphragm system will continue to drop at a rate determined by the reset valve. In this case, the reset valve is set for 1 repeat per minute; that is the diaphragm pressure will continue to drop a specific amount each minute, the exact amount being equal to the unbalance between the two bellows. Remember, the amount of this unbalanced condition is determined by the proportional setting. The reset action, which is an unbalanced condition between the two bellows, continues to drop the diaphragm pressure and open the valve. The pressure in the reset bellows is attempting to equalize with the pressure in the rest of the diaphragm-pressure system but is unable to do so because of the restriction at the reset valve. As the diaphragm pressure falls off, the inner valve opens more and eventually the control variable starts to increase. As it builds, proportional action causes a change in diaphragm pressure that tends to correct the unbalanced condition created by the original drop in the control variable. When the controlled variable builds to the original value, proportional action will have exactly corrected the unbalanced condition between the two bellows. The pressure in both bellows will again be equal but not necessarily at the original 14 lb. Consequently, the beam will be back in its original position. In this type controller, the proportional action only creates an unbalanced condition in the controller when the control variable deviates from the optimum value and provides a reverse balancing correction as the optimum value is restored. The lower the proportional setting or the larger the deviation is, the greater the unbalancing signal will be. Once an unbalanced condition exists, the reset action continues to change the diaphragm pressure, until it strokes the valve completely if necessary, in order to restore the original value. This occurs even for a small deviation.

Rate of Response

Many processes are of such a nature that any deviation of the controlled variable outside certain narrow limits will result in unacceptable products. Any controller equipped with reset action will return the controlled variable to the optimum value, but it can do so only at a specific maximum speed. If there is a sudden drastic change in the controlled variable, reset action will require a considerable period of time to restore the optimum value. In the meantime, the product is lost or ruined. Rate of response partially eliminates this problem by allowing the controller to operate at 0 percent proportional action temporarily. This is accomplished by adding a needle valve in the line supplying the proportional bellows. With this valve partially closed, feedback to the flapper from the bellows system is delayed, allowing the controller to operate at 0 percent proportional action. As the pressure in the proportional bellows gradually changes through the partially closed rate valve, normal proportional and reset action occurs. The rate valve must never be closed more than the reset valve; otherwise the controller will operate at 0 percent proportional action constantly.

Controllers that rely on the restricting orifice, nozzle, and flapper alone to produce the 3- to 15-lb control signal needed to operate a motor valve will generally prove unsatisfactory for anything except the on-off system or those having a very slow change in the controlled variable. Air relays almost invariably form an integral part of the sensitive controller. The back pressure that builds when the flapper moves closer to the nozzle is applied to the air relay that in turn supplies the 3- to 15-lb signal for the motor valve. In many instances, the back pressure at the nozzle will need only to vary 1 lb in order to produce the change from 3 to 15 lb being supplied by the air relay to the motor valve. A no-bleed relay is shown in figure 9.25. The only time the exhaust valve opens is when the diaphragm pressure must be reduced. Other versions of the relay have a throttling valve and vent to atmosphere constantly. The top diaphragm is approximately 12 times as large in area as the lower diaphragm. The two are connected by a metal bracket so that any movement of one causes an equal

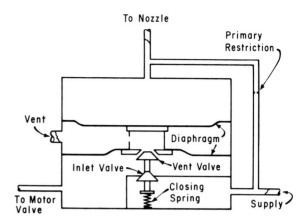

Figure 9.25. No-Bleed Relay for a Diaphragm Motor Valve

movement of the other. An atmospheric vent is connected between the two. The constant 20-lb supply is introduced to the primary restriction and to the valve in the relay. If the nozzle back pressure is increased, the diaphragms will be forced down against the relay valve. Since the valve is rather large, pressure will build on the motor-valve diaphragm quickly. Due to the ratio between the areas of the two diaphragms, approximately 12 times as much pressure will build under the bottom diaphragm before the force exerted will equal the opposing force created by the nozzle back pressure on the top diaphragm. When the two forces are equal, the spring under the relay valve will force the valve closed and trap pressure on top of the motor-valve diaphragm. If the controlled variable later changes so that nozzle back pressure is decreased, the greater pressure under the bottom relay diaphragm will force the assembly up and open the vent. When enough of the pressure has been vented to atmosphere, the forces acting on the relay diaphragms will again be equal and the vent will be closed. The advantage in using a relay is obvious. The motor-valve diaphragm is pressurized and depressurized through large openings rather than through the small controller-restricting orifices. In addition, the nozzle back pressure only changes 1 lb instead of 12 lb to stroke the valve.

The stem in a motor valve passes through a packing gland that prevents gas leakage around the stem. In some installations, this packing will offer substantial resistance to stem movement and the inner valve may not respond to small changes in the control signal. As the name implies, a valve positioner adjusts the valve to the proper opening as determined by the control signal regardless of what pressure it

114

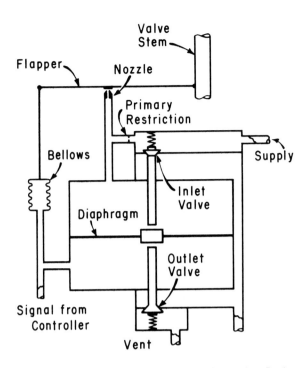

Figure 9.26. Basic Operating Parts of a Valve Positioner

must apply to the motor-valve diaphragm to do so. If, for instance, the control signal is 9 lb (midway between 3 and 15 lb), the inner valve should be half way open. The valve positioner will not be satisfied until the inner valve is in the correct position even though it may have to apply something other than 9 lb to the motor-valve diaphragm in order to get the proper response. Figure 9.26 is a sketch of the basic operating parts of a valve positioner. In an example of a normally open valve, a ¼-lb increase in the control signal applied directly to the motor-valve diaphragm might not be enough to overcome the resistance offered by the packing. The inner valve would not move. However, with a valve positioner, the control signal is applied to a bellows and to the bottom side of a relay diaphragm. The slight increase in control signal causes the bellows to expand and lift one end of the flapper away from a nozzle. The other end of the flapper is attached to the motor-valve stem. As the flapper moves away, the nozzle back pressure will decrease. This back pressure is being applied to the top side of the relay diaphgram, and any decrease allows the diaphragm to move against the supply valve. Opening the supply valve allows pressure to increase on the motor valve diaphragm. The pressure will continue to increase until the stem moves down and brings the flapper back against the nozzle. As the nozzle is closed, the back pressure to the relay

diaphragm will increase until it equals the control-signal pressure under the diaphragm. The diaphragm will move to the center position and allow the relay supply valve to close, trapping the required pressure on the motor-valve diaphragm. A reverse action occurs when the control signal decreases a slight amount. The relay diaphragm moves down, opens the relay vent valve, and vents pressure from the motor-valve diaphragm until the stem is moved to the position called for by the control signal. The valve positioner is usually employed on large motor valves due to the amount of static friction associated with massive packing glands. A high-pressure drop across a valve will occasionally vibrate the inner valve excessively. A valve positioner will aid in stabilizing this vibration.

Electric Controllers

An electric controller valve makes use of a reversible motor. Given a reversible motor, there are numerous methods of converting the rotation of the motor into linear movement suitable for operating a valve. The problem remaining is the control of the motor. Figure 9.27 is a diagram of a controller that will operate the motor if the controlled variable is outside fairly wide limits. A mercury switch acts as a single-pole, double-throw switch capable of assuming an off center position. Almost any type of measuring element is capable of tilting the mercury switch. As an example, suppose the system is used on a valve supplying fluid to a tank where fluid level is the

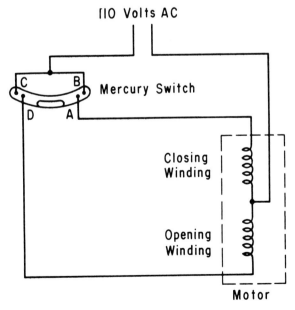

Figure 9.27. Diagram of an Electric Valve Controller

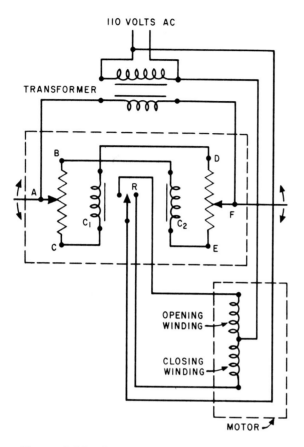

110 VOLTS AC

TRANSFORMER

B

D

A

R

C₁

C₂

F

C

E

OPENING
WINDING

CLOSING
WINDING

MOTOR

Figure 9.28. Circuit Diagram of a Proportional Control System Utilizing a Reversing Motor

controlled variable. If the fluid level increases to the limit adjusted into the controller, the measuring element will tilt the switch enough to make contact between terminals A and B causing the motor to turn the valve in a closing direction. The motor will continue closing the valve until the fluid level drops enough to open terminals A and B. If the fluid level drops too much, the opposite action will occur when contacts C and D close. Obviously, at any fluid level between the two limits, the motor is not energized and the valve opening will remain at a set position. There are many applications for which this controller is not suitable due to the wide allowable deviation of the controlled variable between the two limits.

A circuit diagram of a proportional control system used in conjunction with a reversing motor is shown in figure 9.28. The valve opening in this system is adjusted for each change in the controlled variable, and the valve opening is proportional to the deviation from optimum value. The secondary from a transformer is applied to the movable contacts at terminals A and F. The arm at A is positioned by the

measuring element in the controller and moves each time the controlled variable changes. The arm at F is positioned by the reversing motor and moves each time the motor is energized. These two movable arms are part of two identical variable resistor systems, and their position determines the resistance in each branch of the parallel circuits. The first circuit is from A through C into coil C_1 then through D to F. The second circuit is from A through B into coil C_2 then through E and F. The magnetic force about each coil will be identical, and there will be no movement of the center contact between the coils. If a decrease in the controlled variable drives movable contact A toward B, the total circuit resistance through coil C_2 will decrease, the current through C_2 will increase, the magnetic force around C_2 will increase, and the center contact between the coils will move to terminal R. This energizes the motor circuit that opens the valve. As the valve opens, movable contact F will be driven toward D. This adds resistance to the circuit through C_2 and removes resistance in the circuit through C_1. Eventually, the resistance through the parallel coil circuits will be equal, the current flow through each will be equal, the magnetic field about each will be equal, and the center contact between the coils will move to the off position. A reverse action occurs when the controlled variable deviates in the opposite direction.

Safety Equipment

With the increased activity in offshore production, shutoff equipment, which will stop the flow of gas after damage to the surface equipment, has become an even more critical item. If the surface equipment is damaged or destroyed, it is imperative that the producing well not continue to flow because of economic losses as well as unfavorable public opinion in regard to oil spills, water contamination, and hazards to animal life.

A *velocity-type safety valve* can be installed downhole in the well itself; it consists of a check valve, which is normally held open by spring action. The opening through the unit is properly sized to allow a specific maximum flow without creating an excessive drop in pressure across the unit. If the maximum flow is exceeded, the pressure drop will produce enough force on the check-valve assembly to overcome the spring tension and drive the valve against the seat. The flow of gas stops, and the well remains shut-in until the damage above ground (i.e., broken line, defective control equipment, etc.) has been repaired and the pressure across the check valve equalized. This arrangement is not entirely satisfactory since a ruptured line could conceivably allow some flow to take place but not in the quantity required to close the valve.

The second type of downhole safety unit consists of a *hydraulically operated valve* that is supplied hydraulic pressure through a small supply line from the surface unit. The hydraulic pressure replaces the spring shown in the previous case. If anything happens above ground to cause a loss of hydraulic pressure (i.e., storm, fire, emergency shut-in, etc.), the safety valve will close and remain closed until the pressure across the valve is equalized and hydraulic pressure restored. The advantages of the system are

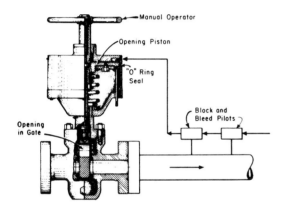

Figure 9.29. Safety Valve (Pilot Pressure Required to Open)

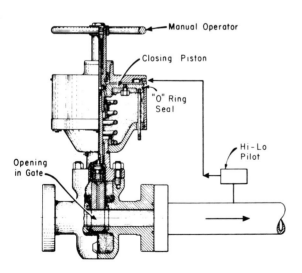

Figure 9.30. Safety Valve (Pilot Pressure Required to Close)

apparent; the valve can be closed at any time from above ground, either intentionally or as a result of damage to equipment.

Above-ground safety valves are generally *piston-operated gate valves,* which may be controlled by pneumatic or hydraulic means. In order to be a truly fail-closed valve, the gate must open as it moves down. Control pressure applied to the piston forces the gate down into its open position. If anything happens to relieve the control pressure such as a broken control line, high or low pressure, and so forth, the valve will close due to the closing effect of the coil spring (fig. 9.29), and the pressure inside the valve against the stem forces the gate up into the closed position. A similar valve is shown in figure 9.30. However, this valve opens as the gate moves up. Control pressure must be applied to the piston to close the valve. This control pressure is applied when a high or low pressure exists in the piping system. A major disadvantage of this system is the inability of the valve to close if any of the surface equipment is damaged by a storm or fire.

The pilot controllers required to operate the valve in figure 9.29 are shown in figures 9.31 and 9.32. One supply pressure is used, and it can be passed through any number of pilots required to monitor the system. These pilots are connected in series and normally apply full supply pressure to the safety valve to hold it in the open position. If any one of the pilots operate due to high pressure, low pressure, or emergency shutdown, the supply pressure is blocked,

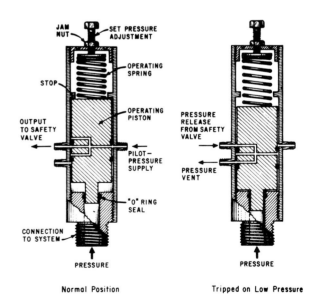

Figure 9.31. Low-Pressure Block-and-Bleed Pilot

117

the pressure on the closing piston is bled to atmosphere, and the safety valve closes. Some pilot valves will remain in this blocked condition until manually reset, others will automatically reset and open the safety valve when the abnormal pressure condition is corrected. Due to their operating characteristics, these pilots are often referred to as block-and-bleed units. A second type of pilot is shown in figure 9.33. The operation of this unit is exactly reversed since it blocks the control pressure to the closing piston on the safety valve until an abnormal condition exists in the system. When the abnormal condition occurs, the pilot moves to the open position and applies control pressure to the closing piston of the safety shutoff valve.

Most vessels used between the wellhead and the transmission system are equipped with *pressure-relief valves* to prevent overpressure. One such device is

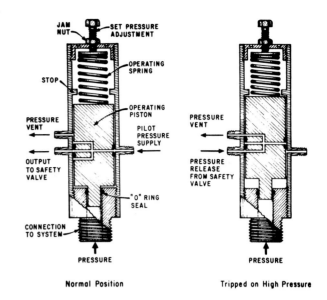

Figure 9.32. High-Pressure Block-and-Bleed Pilot

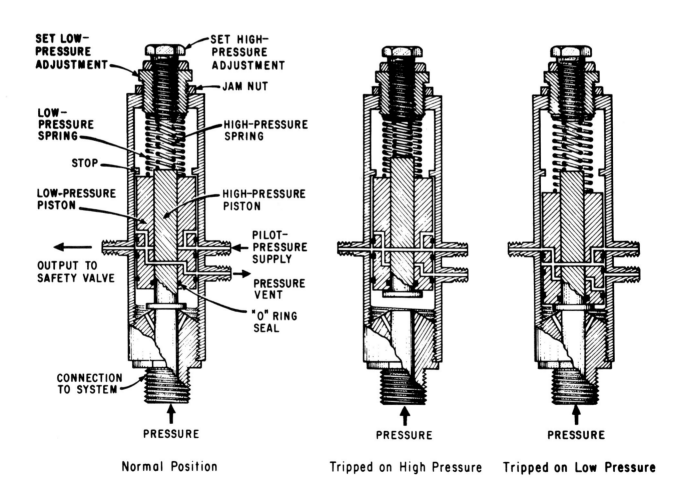

Figure 9.33. Combination Pilot for Safety Valve Requiring Pressure to Close

118

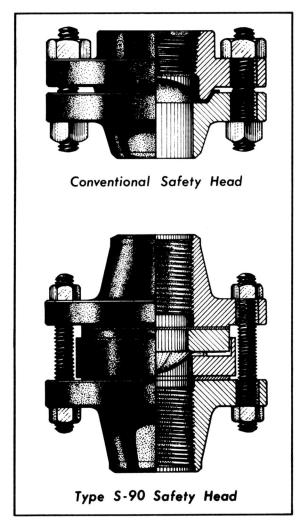

Conventional Safety Head

Type S-90 Safety Head

Figure 9.34. Bursting Disc Pressure-Release Devices

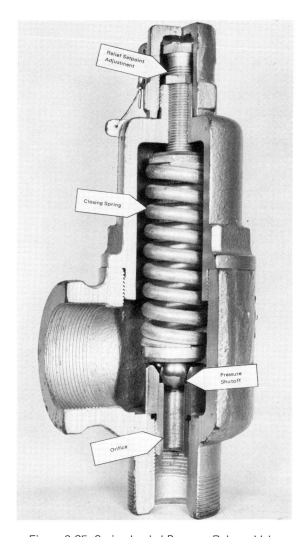

Figure 9.35. Spring-loaded Pressure-Release Valve

shown in figure 9.34 and consists of a metal diaphragm between a set of flanges. The diaphragm can be designed to rupture at any desired pressure and vent excess gas to atmosphere. However, once ruptured, the diaphragm must be replaced before the vessel can be put back on-stream.

A spring-closed relief valve is shown in figure 9.35. The spring tension can be adjusted to hold the valve against the orifice until a predetermined pressure is reached. At that time, the force exerted by the pressure exceeds the spring tension and the valve is lifted off the orifice venting excess gas. The major problem with this unit is leakage. As the set pressure is approached, the valve tends to leak; consequently,

the normal operating pressure in the system should not be above 90 percent of the set point on the relief valve. In addition, the valve and orifice are generally damaged with each relieving action and must be resurfaced to prevent leakage during normal pressure conditions.

Pilot-operated relief valves are used when normal operating pressures are very near the relief set point and when a specific decrease (blowdown) in pressure is required following relief action. Such requirements are found mostly in refineries, gas plants, and so forth, and relief valves of this type are almost never needed in field operations.

X

MEASUREMENT OF NATURAL GAS AND GAS LIQUIDS

Natural gas is a valuable asset and a superior fuel. Good business practices dictate the need for efficient handling of natural gas and natural gas liquids. Efficient handling of these products will include good measurement practices.

The volume of gas is the fundamental basis for settlement in most gas-sales transactions. Payments for royalties and taxes are usually based on measured volumes. Reports to state regulatory bodies and to the Federal Power Commission, when required, are based on volumes measured in the field. Gas, being a vapor, is not subject to conventional methods of storage in large quantities. Therefore, it must be measured instantaneously as it flows through a pipeline. Natural gas liquids may be measured as they flow through a line, or they may be stored and measured in the storage vessel.

Metering of fluids may be described as the practice of measuring volumes or rates of flow by actually passing the fluids through some type of meter. While metering is the most common method of determining volumes or rates of flow, there are other methods of measurement that are sometimes used. For gas, these would include calculation of the volume in a pipeline (i.e., line pack) or in storage on the basis of the line or storage volume and pressure; application of pipeline flow formulas; by estimation from time and rate of usage factors; by computation of compressed volumes from temperature, pressure, cylinder displacement, and compressor speed; and by other similar techniques that may be practical. For liquids, the gauging of storage tanks would be included.

A fluid flowing through a line can be measured by placing a constriction in the line to cause the pressure of the flowing fluid to drop as it passes the

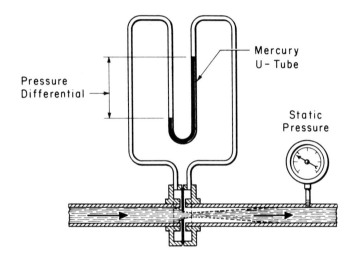

Figure 10.1. Diagram of the Pressure Drop across an Orifice Plate

constriction. This pressure drop is called *differential pressure* (fig. 10.1). There is a direct relationship between the rate of flow and the amount of this pressure drop, or differential. This principle has been widely used and has been developed into a precise and accurate means of measuring fluids when all factors are taken into consideration and when ideal conditions for its use prevail. Because this principle is so widely used, the method that makes use of the principle known as orifice measurement will be given emphasis herein.

Both gas and liquids may be measured by use of other measurement techniques including positive displacement meters, turbine meters, venturi meters, flow nozzles, critical flow provers, elbow meters, variable area meters (rotameter), and others. The selection of the measurement method to be used should be made only after careful analysis of several factors including the following:

1. accuracy desired;
2. expected useful life of the measuring device;
3. range of flow, temperature, and so forth;
4. maintenance requirements;
5. power availability, if required;
6. liquid or gas;
7. cost of operation;
8. initial cost;
9. availability of parts;
10. acceptability by others involved;
11. purpose for which measurements are to be used; and
12. susceptibility to theft or vandalism.

NATURAL GAS MEASUREMENT

Measurement of gas requires the unit of volume to be defined. In other words, the unit, temperature base, pressure base, and other factors must be specified or determined. There are currently no universally accepted temperature and pressure bases for cubic foot measurements although this may eventually come about. The bases most likely to find acceptance are 60 F (520 R) and 14.73 psia. A cubic foot of gas, when such is the unit, may be defined as being the amount of gas required to fill a cubic foot of space when at a temperature of 60 F and under a pressure of 14.73 psia. Gas may be measured by an orifice meter in other units such as the pound, therm, MMBtu, and so forth. These, too, must be defined.

Orifice Meters

The orifice meter actually consists of several parts. These parts make up what are usually referred to as the primary element—(i.e., meter tube, etc.) and the secondary element (the recorder). Since gas is a highly compressible fluid, its density will vary considerably with the pressure existing at the constriction or orifice plate. Consequently when gas is measured with an orifice meter, it is necessary to measure both the differential pressure and the flowing pressure, which in gas-measurement language is called the static pressure. There are several other measurements and factors to be taken into account in the accurate measurement of gas such as temperature and specific gravity.

Primary Element

The meter run, meter fittings or flange unions, and the orifice plate, which make up the primary

element of a meter setting, should be a shop-fabricated unit manufactured by a reputable manufacturer of such equipment. The unit should be manufactured in accordance with the latest edition or revision of the American Gas Association's Gas Measurement Committee Report No. 3 (revised, 1969). The recommended standards on gas measurement expressed in Report No. 3 are the result of several years of intensive research and experimentation by special joint committees of the American Gas Association (AGA) and the American Society of Mechanical Engineers (ASME). The original report was approved for publication by the director of the National Bureau of Standards. Every revision has resulted from data collected during careful testing, and each revision has served to improve the utility of the recommendations published.

When a primary element is constructed in accordance with Report No. 3, the flow of gas through it will be streamlined and there will be no obstructions within to disturb the flow pattern and contribute to inaccurate measurement. It is advisable to make direct reference to AGA Report No. 3 when planning or constructing a meter run. Report No. 3 and its specifications and recommendations relate to and should be understood to be limited to the following two types of orifice meters:

1. orifice meters with circular orifices placed concentrically in the pipeline and having upstream and downstream pressure taps as specified for flange taps and
2. orifice meters with circular orifices located concentrically in the pipeline and having upstream and downstream pressure taps as specified for pipe taps.

Report No. 3 covers the measurement of flowing fluids, particularly natural gas, as related to the primary element and to the methods of calculation. The report does not cover equipment used in the determination of the pressures, temperatures, and other variables that must be known for accurate measurement. The installation sketch figure 10.2 shows a basic meter-tube length. This length meter tube will accomodate a restriction in the pipeline upstream of the orifice. Meter-tube lengths may be reduced under specific circumstances.

The many experiments made in development of the orifice flow constants and other factors given in AGA Report No. 3 involved the use of long, straight lengths of pipe both upstream and downstream of the orifice. Many of these experiments involved the use of straightening vanes. These are usually bundles of

Installation Sketch Fig. 3-A
AGA Report No. 3

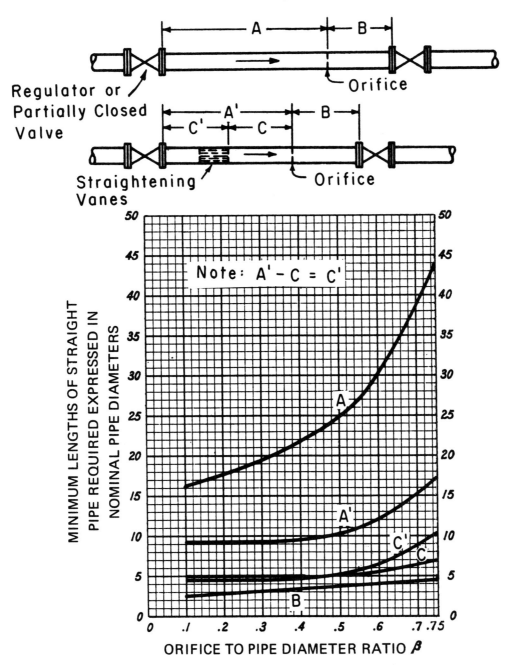

NOTE 1—When "Pipe Taps" are used, lengths A, A', and C shall be increased by 2 pipe diameters, and B by 8 pipe diameters.

NOTE 2—When the diameter of orifice may require changing to meet different conditions, the lengths of straight pipe should be those required for the maximum orifice to pipe diameter ratio that may be used.

Figure 10.2. Installation Sketch for a Basic Orifice Meter Tube

122

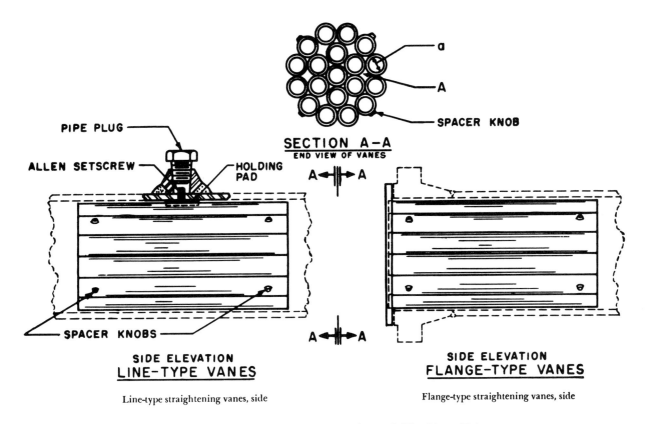

SECTION A-A
END VIEW OF VANES

PIPE PLUG

ALLEN SETSCREW

HOLDING PAD

SPACER KNOB

SPACER KNOBS

SIDE ELEVATION
LINE-TYPE VANES

Line-type straightening vanes, side

SIDE ELEVATION
FLANGE-TYPE VANES

Flange-type straightening vanes, side

Figure 10.3. Straightening Vanes for an Orifice Meter Tube

small diameter tubing tack welded together in a concentric pattern and placed in the upstream section of the meter run as indicated previously. Straightening vanes are as shown in figure 10.3. Since most field applications at some time involve conditions reflected in the installation sketch shown in figure 10.2, it is recommended that field installations meet the requirements portrayed as to upstream and downstream lengths as well as to the use of straightening vanes. When space or other factors make it desirable to shorten upstream or downstream lengths, installation sketches shown in the AGA Report should be referred to.

The installation of straightening vanes will reduce considerably the amount of straight pipe required upstream of the orifice. Vanes should be reamed as thin as practical at both ends and should be arranged so they are symmetrical with the smallest diameters in the center of the bundle, though the tubes need not be the same size. The purpose of the vanes is to eliminate swirls and crosscurrents set up by the pipe fittings and valves preceding the meter tube.

The orifice plate is held in position in the meter run by conventional orifice flanges or by a special orifice fitting. Orifice fittings are constructed to permit easy changing and inspection of orifice plates. Since such inspections where conventional flanges are used may be quite difficult, fittings are now generally used. Some fittings are designed so that plates can be changed without shutting off the flow of gas (fig. 10.4); in others, the gas must be shut off or the meter run by-passed. The circumstances peculiar to a given measuring situation govern the decision as to the type of installation to be made.

Secondary Element (Recorder)

Orifice recorders normally include devices for measuring both differential and flowing pressures. Three types of differential-measuring devices are generally used. These are the mercury manometer (fig. 10.5), the bellows (fig. 10.6), and the force-balance type (fig. 10.7). Some flow or orifice calculations for measurement of gas will have one or more factors that will be affected by the type of meter used. AGA Report No. 3 lists such factors for mercury and bellows meters. Force-balance type transmitters will use the factors listed for bellows

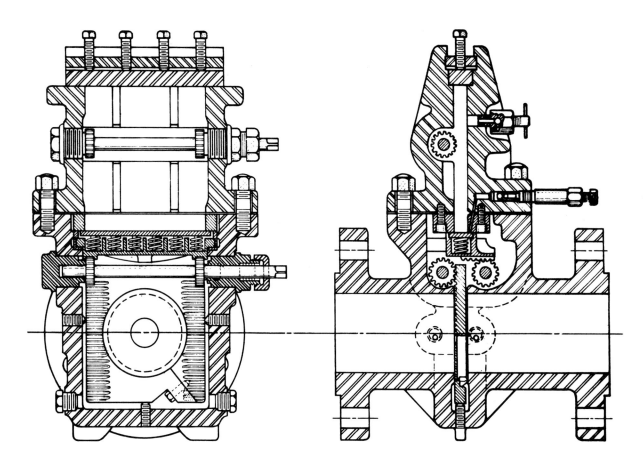

Figure 10.4. Bypass-Type Orifice Flange Fitting

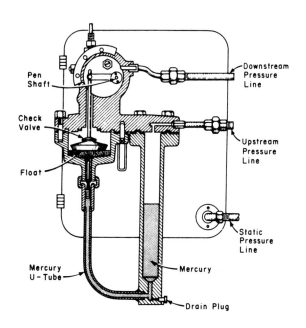

Figure 10.5 Mercury-Manometer-Type Differential-Pressure Measuring Device

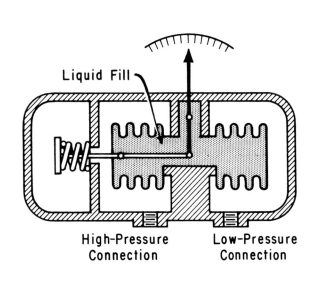

Figure 10.6. Schematic View of Bellows-Type Differential-Pressure Measuring Device

124

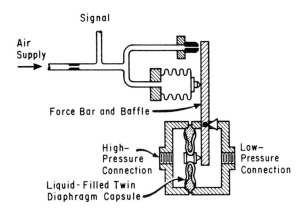

Figure 10.7. Force-Balance Differential-Gauge Differential-Pressure Measuring Device (Schematic)

meters. Flowing (or static) pressures are measured by ordinary pressure recorders similar to those described in a previous chapter.

For gas service, a commonly used rule states that the differential range of the meter in inches of water should not exceed the static pressure in pounds per square inch absolute. That rule may require modification to conform to the characteristics of a particular application. Most gas measurements are made by use of 0- to 10-in., 0- to 20-in., 0- to 50-in., or 0- to 100-in. differential-range meters. Common practice is to select a static range that will permit recording in the upper half of chart. For example, use 0- to 1,000-lb range for 700-lb or even 900-lb static pressure.

Temperature Determination

Thermometer wells should be placed downstream of the orifice plate not nearer the plate than dimension B in figure 10.2. If it becomes necessary for the well to be placed upstream the orifice plate, it should be upstream of dimension A or A'. If indicating thermometers are used, they should be protected against breakage and the readings should be observed and posted on meter charts as frequently as may be practical but at least twice each chart period; that is, when the chart is placed and when it is removed. Recording thermometers should be used on most gas-sales meters. Common types are described in chapter nine. Recorder temperatures during periods of no flow should be disregarded in volume computations.

Specific Gravity Determination

The most commonly used method for the field determination of the specific gravity of a gas is by some form of weighing such as the Edwards Balance. This instrument compares the density of the gas being tested with the density of air. The calculation of specific gravity is based on the principle that the densities of two gases at base pressure are in inverse ratio to the pressures that provide equal buoyant forces for the two gases. This method of specific gravity determination is used at locations involving both small and large volume sales and purchases of gas.

The recommended equipment for use in obtaining gas gravities by the balance method is as follows:

1. the Acme Senior Gas Gravity Balance (fig. 10.8), for use by measurement specialists and at all locations where large volumes of gas are tested and

2. the Acme Junior Gas Gravity Balance, for use by measurement specialists working with small volumes of casing-head gas.

Two recognized technical standards are listed for reference. The first is recommended for use.

1. "Standard Method for Determining the Specific Gravity of Gases," Natural Gas Processors Association of America's Standard 3130

2. "Tentative Standard Methods for Determining the Specific Gravity of Gases," W.G.P. & O.R. Association's TS-391

The proportional torque differential indicator operates on the principal that the torque produced on the shaft of an impulse wheel activated by a jet of gas is proportional to the specific weight of the gas. The difference in torque produced by air and the tested

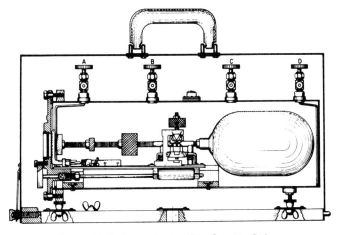

Figure 10.8. Acme Senior Gas Gravity Balance

125

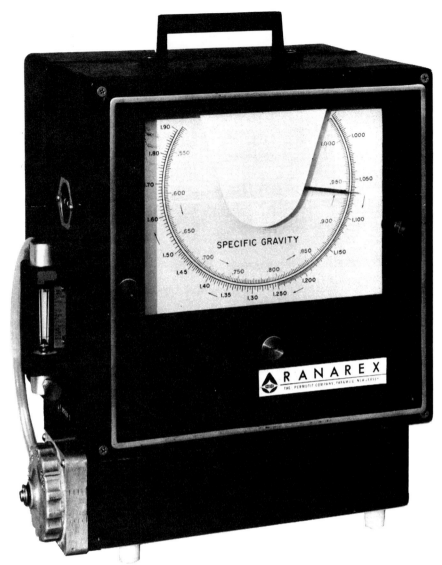

Figure 10.9. Proportional Torque Differential Indicator for Measuring the Specific Gravity of Gas

gas is shown on an indicator scaled to read in terms of specific gravity. Recommended equipment is the Portable Ranarex Indicator as illustrated in figure 10.9. This indicator is for determining specific gravities applicable to allocation meters.

Recording gravitometers may be used when a continuous record is desirable or required. The types most frequently used are those making use of weighing methods (fig. 10.10) and those using the momentum method.

Meter Piping

The differential lead lines that connect the instrument to the orifice flange or pipe may be ¼-in. nominal diameter pipe for lengths up to 15 ft and ½-in. nominal diameter pipe for lengths up to 40 ft. Tubing of equivalent internal diameter may be used in place of pipe. Distances greater than 40 ft are feasible but are not recommended. Where such distances are necessary, either pneumatic or electric transmission should be considered. Both the upstream and downstream orifice connections should be made from the same side of the flange or fitting, preferably from the top (with a fitting placed on its side). In no case is it recommended that differential connections be made from the bottom of the fitting, flange, or line. Differential lines should be carried together with a slope as great as possible but not less than 1 in./ft. If the orifice is above the meter, the lines should slope up to the orifice. If the orifice is below the meter, the lines should slope down to the orifice.

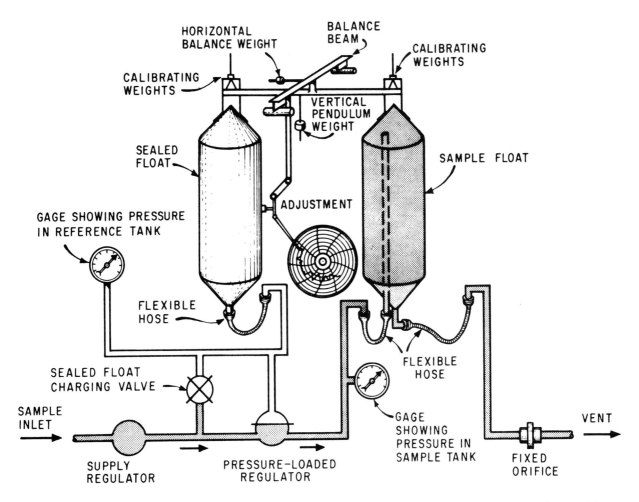

Figure 10.10. Schematic Diagram of a Recording Gravitometer Having Gas-filled Drums Suspended from a Balance Beam

Measurement Calculations

Basic Flow Equation

In measuring the flow of gas with orifice meters, the flowing or static pressure of the gas as well as the differential pressure across the orifice must be obtained. Their relationship to rate of flow is expressed by the following equation:

$$Q_h = C'\sqrt{h_w p_f} = \text{Rate of Gas Flow, cu ft/hr at Base Conditions}$$

Where

C' = Orifice Flow Constant, Corrected for Operating and Base Conditions

h_w = Differential Pressure across Orifice, in. of water

p_f = Static Pressure, psia

The product of the square roots of the differential and static pressure absolute $\sqrt{h_w p_f}$ is commonly referred to as the pressure extension. Pressure extension tables are available in book form as a convenience in calculating gas flows.

Orifice Flow Constant

The orifice flow constant, C', is obtained by multiplying a basic orifice flow factor, F_b, by various correcting factors that are determined by the operating conditions, contract requirements, and physical nature of the installation. This is expressed in the following equation:

$$C' = (F_b)(F_{pb})(F_{tb})(F_g)(F_{tf})(F_r)(Y)(F_{pv})$$
$$(F_m)(F_{s1})(F_e)$$

127

Where

F_b = Basic Orifice Flow Factor, cu ft/hr

F_{pb} = Contract Pressure Base

F_{tb} = Contract Temperature Base

F_g = Specific Gravity Factor

F_{tf} = Flowing Temperature Factor

F_r = Reynolds's Number (Viscosity) Factor

Y = Expansion Factor

    Y_1 = Based on the Upstream Static Pressure

    Y_2 = Based on the Downstream Static Pressure

    Y_m = Based on a Mean of the Upstream and Downstream Static Pressures

F_{pv} = Supercompressibility Factor

F_m = Manometer Factor for Mercury Meter

F_{s1} = Seal Factor, Applies Only to Mercury Meter Sealed with a Liquid

F_e = Orifice Plate Expansion Factor

All orifice-meter factors or coefficients should be selected from or calculated by use of AGA Report No. 3 unless the use of another method is required. A definite procedure for the calculation of coefficients can be established to insure that all such coefficients will be calculated in the same manner and that, starting with the same basic data, any two persons using this procedure will arrive at identical answers for each coefficient.

Most natural gas streams contain water, and some of these streams are saturated with water. Errors in gas measurement and in gas composition may result if the presence of water vapor in the gas is not properly accounted for. The errors in volume measurement will be insignificant at pressures above 200 psig, but at lower pressures the errors can be substantial. The same is true of errors in composition as determined by analysis, although these errors are several times greater than the errors in volume.

Measurement error will not usually result because of water vapor in the gas if a field-determined specific gravity is used in the gas volume computation and if this gravity was determined on the wet sample. However, when a specific gravity is calculated from an analysis of the gas, these analyses are customarily reported on a dry basis and must be converted to a wet basis before a wet specific gravity can be obtained. Once the water content is determined, the volume of water obtained may be subtracted from the wet volume of gas to arrive at the gas volume expressed on a dry basis.

Accuracy of Measurements

While it is impractical to list all the various factors that may affect the accuracy of measurement obtainable with differential-type flow instruments, it will perhaps be of some value to list the more common factors as sources of constant errors and variable errors. This is an arbitrary classification because, under certain circumstances, the variable errors may result in constant or nearly constant errors whereas several of the errors listed as constant may prove to be variable. The purpose of the lists is to provide a summary of possible sources of errors that may be helpful in locating and correcting errors and thereby obtain better measurement. In most cases, corrections can be made easily mechanically or through adequate maintenance. Special situations such as freezing and pulsation are discussed in more detail.

References may be found that indicate the relative effect or amount of error caused by many of these factors. However, it is not recommended that such data be used as a basis for applying corrections to flow calculations where these factors are present or suspected.

Flow measurement is often used as a basis for control only. As a general rule, the more accurate measuring installations will give more accurate control. However, in most cases, satisfactory control can be expected as long as the errors remain constant.

The constant errors include the following:

1. incorrect information as to the bore of the orifice plate;
2. contour of the orifice plate (convex or concave);
3. dullness of the orifice edge;
4. thickness of the orifice edge;
5. eccentricity of the orifice bore in relation to the pipe bore;
6. incorrect information as to the pipe bore;
7. excessive recess between the end of pipe and the face of orifice plate; and
8. excessive pipe roughness.

The variable errors include the following:

1. flow disturbances caused by insufficient length of meter tube or irregularities in the pipe, welding, and so forth;
2. incorrect locations of differential taps in relation to the orifice plate;
3. pulsating flow;

4. progressive buildup of solids, dirt, sediment and so forth on the upstream side of orifice plate;
5. improper check-valve operation;
6. accumulation of liquid in the bottom of a horizontal run;
7. liquids in the piping or meter body;
8. changes in operating conditions from those used in the coefficient calculations (i.e., specific gravity, atmospheric pressure, temperature, etc.);
9. incorrect zero adjustment of the meter;
10. nonuniform calibration characteristic of the meter;
11. corrosion or deposits in the meter range tube or float chamber;
12. emulsification of liquids with mercury;
13. dirty mercury;
14. incorrect arc for meter pens;
15. formation of hydrates in meter piping or meter body;
16. leakage around the orifice plate (applies to orifice fittings);
17. wrong range on chart;
18. incorrect time for rotation of chart;
19. excessive friction in the meter's stuffing box;
20. meter not level (mercury-type only);
21. excessive friction between the pen and chart; and
22. overdampening of the meter response.

Processing Meter Charts

Charts should first pass through the hands of the operating location staff and, when necessary, a measurement specialist who will note on the back of the chart the gravity, temperature, supercompressibility factor, and any other information affecting the volume and will verify the field data imprinted on the back of the chart. On the basis of past history or familiarity with conditions at the meter, the measurement specialist will estimate volumes when the meter was inoperative because of a stopped clock, pen not marking, freezing, overranging, mercury blown, and so forth.

When it becomes necessary to estimate volumes because of an inoperative meter, contract provisions governing such estimations should be followed.

Many sales contracts include a provision to the effect that if for any reason meters are out of service so that the amount of gas cannot be determined from chart computations, the gas delivered during the period in which the meter was inoperative shall be determined by using the first of the following methods that is feasible.

1. By using check meter readings.
2. By correcting the error with mathematical calculations if such error is ascertainable by calibration or test. This includes a wrong-sized orifice plate, a plate placed in backwards, a differential pen not zeroed, a static pen not calibrated, a mercury level that is high or low, and so forth.
3. By estimating the volume by comparison with deliveries during a period when the meter was operating properly.

Testing and Maintenance

In making a routine meter test, the following items may be included. A portable deadweight tester or test pressure gauge of known accuracy may be used to calibrate a static pressure element. The check should be made at high, low, and medium chart range.

Meter lines should be inspected for leaks. Obstructions in the piping other than hydrates may sometimes be removed by use of a solvent. Specific instructions for testing the differential calibration are included in the operating instructions booklet furnished with new meters. These procedures provide for comparison of the recording to a water manometer as a standard when equal pressure is exerted on the recorder and the manometer.

For most replacement purposes and usually with new meters, spring-driven standard clocks are used. However, battery- and electric-driven clocks are becoming more common. Normally, a spring-driven clock will run fast during the winter and slow during the summer due to the expansion and contraction of metal parts. At the approach of cold weather, clocks should be cleaned. Care should be taken that clock-mounting screws are not too tight causing the clock to run slow. The clock shaft should be protected carefully during cleaning operations. A bent clock shaft will not rotate the chart concentrically.

Differential and static recording pens should be cleaned, inked, and the time-arc checked. The static pen should be adjusted to record behind (i.e., lag) the differential pen approximately 15 minutes on 24-hour charts and 2 hours on 7-day charts. This is to make recordings compatible with chart-integrating devices and facilitate integration of the charts.

Mercury-Type Meters

When it becomes necessary to replace a worn differential shaft, the whole stuffing box and shaft should be replaced with a Teflon-bearing stuffing box. The allowable stricture of the differential shaft should be not more than 0.4 in. of water for high-pressure meters and 0.1 in. for low-pressure meters. If this frictional tolerance is exceeded, remove the stuffing box and lap the shaft to the bearing with jeweler's rouge. Care must be taken that all rouge is removed, by several washings in solvent, after the lapping is complete. The stuffing-box lubricant should vary with the type of service and climatic conditions. Overpacking with grease may cause the shaft to bind.

The mercury must be clean. The mercury chamber should first be cleaned with an approved type of solvent. The mercury may be cleaned by straining through a chamois or coarse cloth or washing in a 10 to 20 percent solution of nitric acid and rinsing with clear water.

The float must be properly centered and should travel freely. Erratic float travel may be due to a worn pin or to the manometer not being level.

See that the check valve (inside meter bodies) operates freely.

It is extremely important that the manometer be level. If the instrument level is broken, a 6-in. machinist's level should be used across the top of one of the mercury chambers that has had the cover removed.

Bellows-Type Meters

Check for evidence of leakage of meter internal fill liquid. Where leaks occur, it is not recommended that the repairs and refilling be attempted in the field but that the faulty bellows unit be exchanged for a new or factory-overhauled unit.

Check for liquid and/or foreign material accumulations in the bellows housing. Overrange and underrange the bellows, and check for sticking overrange seal valves.

Check for evidence of internal and external corrosion that may affect the meter operation.

Check for excessive meter dampening.

Orifice Plate

Orifice plates should be inspected to ascertain that—

1. The upstream edge is sharp and without a wire edge.
2. The face of the plate is flat.
3. The face of the plate is smooth and without pits.
4. The correct size of the bore, measured by micrometer, is stamped on the plate.
5. No dirt or ice has collected against the orifice plate.

Orifice Fittings

If the meter tube is equipped with an orifice fitting, the following observations and operations should be made.

1. Packed glands must be kept tight.
2. Lubricate the moving parts.
3. Remove the plate carrier for inspection on a routine schedule.
4. Actuate the moving parts to prevent them from becoming frozen.
5. See that the orifice plate holder is not frozen into place or that water does not collect about the isolating valve and burst the upper chamber during very cold weather.

Measurement Problems

Freezing

Hydrate formations at the orifice, in meter piping, or in the meter chamber may occur when the temperature of the wet gas being measured falls below the hydrate temperature. The chart, on which recordings have been made while the meter was partially frozen, should have a full explanation written on the face of the chart and estimated static and differential should be lines drawn in. Preventive measures include—

1. elimination of piping leaks;
2. installation of line heaters;
3. installation of a heated meter house;
4. dehydration of gas;
5. use of inhibitors;
6. enlarging the meter piping and valves to ½ in. maximum; and
7. replacing needle valves with plug or gate valves.

Pulsating Flow

Pulsations in a pipeline originating from a reciprocating system or some other similar source consist of sudden changes in both the velocity and pressure of the flowing fluid. The pressure changes are the more apparent and resemble low-frequency sound waves traveling in the flowing medium with a velocity independent of the velocity of the flowing fluid. The most common sources of pulsation involved in gas measurement are—

1. reciprocating compressors, engines, or impeller-type boosters;
2. pumping or improperly sized pressure regulators and loose or worn valves;
3. irregular movement of quantities of water or oil condensate in the line; and
4. intermitters on wells and automatic drips.

Reliable measurements of a gas flow with an orifice meter cannot be obtained when appreciable pulsations from any source are present at the point of measurement. No way has been found to determine or predict correction factors to compensate for such errors.

In order to obtain reliable measurements, it is necessary to suppress the pulsations. In general, the following items are valuable in diminishing pulsation and/or its effect on orifice flow measurement.

1. Locate the meter tube in a more favorable location with regard to the source of pulsation such as at the inlet side of regulators or increase the distance from the source of pulsation, and so forth.
2. Insert capacity (volume), restriction, or specially designed filters in the line between the source of pulsation and the meter tube in order to reduce the amplitude of the pulsation.
3. Operate at differentials as high as is practicable by replacing the orifice plate in use with an orifice plate having a smaller orifice or by concentrating flow in a multiple tube installation through a limited number of tubes.
4. Use smaller sized tubes and keep essentially the same size of orifice while still maintaining the highest practical limit on the differential.

As yet no instrument has been developed that will give precise quantitative effects of pulsation on flow measurement. Instruments have been developed, both mechanical and electrical in nature, that indicate the presence of pulsation that could affect the measurement.

Slugging

The conditions commonly called slugging refer to a liquid (water, oil, or condensate) accumulation in a gas line. In low-pressure lines, the liquid will gather at a low place in the line and restrict the passage of gas until enough gas pressure has accumulated to blow through the liquid. In the high-pressure system, the liquid will sweep up to and through the orifice. Both conditions produce erratic recordings and inaccurate measurement of an undeterminable extent. The use of drips or liquid accumulators of various types is often the only feasible solution to the problem. When chokes are used on wet-gas lines, they should be placed downstream of the orifice.

Sour Gas

Freezing and corrosion are two frequent problems when measuring gas containing hydrogen sulfide. Corrosion in a closed line free of air and water is negligible, and most meter corrosion is due to hydrogen sulfide in the surrounding atmosphere. Possible remedies include the following changes.

1. Static spring—316 Stainless steel is generally satisfactory.
2. Differential pen shaft—use Teflon bearings that are unaffected by hydrogen sulfide. Lubrication with a silicone lubricant is helpful.
3. Pen—self-feeding pens give better service as they are more closely sealed against the atmosphere.
4. Clocks—vaporproof clocks are essential. The rubber seal should be coated with varnish. The winding stem and chart hub stem should be coated with grease.
5. Seal pots—seal pots are essential for protection of mercury in mercury-type meters. Some recommended sealing fluids are ethylene glycol or glycol-base antifreeze compound with 40 percent water. Add 4 ml of 25 percent formaldehyde per gal as an inhibitor. The orifice factor must be properly corrected when using sealing fluids.

Other Methods of Gas Measurement

Displacement Meters

Displacement meters are often referred to as positive displacement meters since they afford a positive volume in cubic feet at flowing conditions regardless of the flowing temperature or the specific gravity of the gas. Displacement meters used in the field are generally of two types: the rotary or impeller type and the slide-valve diaphragm type. Turbine meters are sometimes included in this category.

The rotary and diaphragm types of displacement meters contain measuring elements of known volume and valve arrangements to channel the gas into and out of the measuring elements. They also are equipped with a counter or index dial to count the number of times the volume container or element may have been filled during the measurement process.

Since displacement meters measure the volumetric displacement of the gas at flowing conditions, it usually becomes necessary to record the pressure or provide some other means of adjusting the meter displacement to the base pressure of the defined measurement unit. Likewise, when required, flowing temperature adjustments to the base temperature should be made.

Depending on the class of service, several types of volume recording and/or totalizing register devices may be installed on displacement meters. These include the following devices.

1. The dial index is used when constant pressure can be maintained on meter.
2. The dial index and recording pressure gauge is used when constant pressure cannot be maintained on meter. Chart pens record the gauge pressure on the calibrated portion of the chart and the volume on the outer edge of the chart. The latter registration is by means of a series of waves, or loops; each complete wave or loop indicates that one (or ten) Mcf has passed through the meter at the pressure indicated by the pressure pen. Volumes computed by counting the loops will check very closely with the dial-index reading.
3. The base pressure corrector is used to correct the volume of gas passing through the meter to the base pressure. This volume is registered on an accumulating counter. No pressure correction is needed for calculating the volume recorded.

Volumes as recorded by positive meters are independent of specific gravity correction. For a dial index and recording pressure gauge, the recorded volume must be corrected for pressure and, if necessary, for temperature. The base pressure corrector needs only the temperature correction.

It is recommended that field personnel not attempt field repairs on displacement meters although repairing and replacements may be made in the field by trained personnel with proper tools. It will be sufficient that measurement personnel check to see that the meter is operating within its rated capacity since increased speed means increased friction and wear. Of the several standard methods of testing displacement meters in the field, it is recommended

that a critical flow prover be used when the gas pressure can be maintained above 15 psig.

Mass-Flow Meter

The orifice meter may be used to measure gas on a mass-flow basis into pound units. The technique requires determination of the density of the gas being measured and the differential pressure across the orifice. Equipment is now available to measure the density of a flowing gas. The equipment is called a densitometer and is available for both electric and pneumatic applications. In the use of the orifice meter for making mass measurements, the densitometer is substituted for the static-pressure element and makes determination of the specific gravity and supercompressibility corrections unnecessary.

The following equation from AGA Report No. 3 may be used for mass flow computations:

$$W = 1.0618 \, F_b \, F_r \, Y \, \sqrt{h_w \, \gamma}$$

Where

W = Flow Rate, lb/hr

F_b = Basic Orifice Factor

F_r = Reynolds Number Factor

Y = Expansion Factor

h_w = Differential Pressure

γ = Specific Weight of Gas at Flowing Conditions, lb/cu ft

Another mass meter is a volumetric device called the Vortex-Velocity mass-flow meter where weight flow is determined from the volume flow in cubic feet at flowing conditions and the specific weight in pounds per cubic foot.

Still a third mass-flow meter makes use of a specially designed gyroscope that is caused to precess around its major axis when the torque from the turbine is applied to its minor axis. The speed of the precession is directly proportional to the weight-flow rate. The cumulative mass flow is recorded on a counter operated by the precession.

Turbine Meter

Several makes and designs of turbine meters are currently available. These make use of the flowing gas as a force imparted to a bladed rotor. By use of appropriate gearing, revolutions of the rotor may be converted to volume. Accuracy curves are usually developed for each turbine meter and proving or calibration techniques are being developed. Filters ahead of turbine meters are almost a necessity to permit sustained accuracy and trouble-free operation. Progress in this type of measurement indicates many turbine meters may soon be in use.

Elbow Meter

Centrifugal force in the curve of a pipe elbow can be used to measure flow. For accuracy, the elbow should be calibrated using some other acceptable measurement as a standard. Accuracy is not usually the objective when elbow meters are used. Relatively little pressure loss or differential pressure is created. Because of this, the meters are used primarily for control or other operation purposes.

NATURAL GAS LIQUIDS MEASUREMENT

Orifice Meter

Field measurements of natural gas liquids are accomplished by the conventional gauging of tanks and by use of various metering techniques. The orifice meter is sometimes used. Installation and operation requirements are about the same as for gas. Some of the basic data are used in the calculation of orifice meter recordings into pound or gallon units as are used for gas. However, there are different factors to be applied.

For measurement into gallons, for example, the American Meter Company's *Orifice Meter Constants: Handbook E-2* may be used. The equation for this handbook is as follows:

$$Q_h = C' \sqrt{h_w}$$

Where

Q_h = Rate of Liquid Flow, gal/hr

C' = Orifice Constant (F_b X F_{gt} X F_{sl} X F_r)

h_w = Differential Pressure, in. of water

F_b = Orifice Factor

F_{gt} = Specific Gravity Factor

F_{sl} = Factor for Seals, if Required

F_r = Reynolds Number Factor

For determination into pound units the following equation from *Principles and Practices of Flow Meter Engineering*, 8th edition, published by Foxboro Company in 1961, may be used:

$$W = SND^2 F_a F_m F_c F_p \sqrt{G_f h_w}$$

Where

W = Rate of flow given in pounds per twenty-four (24) hours.

S = A value determined for the bore of the orifice and the internal diameter of the metering tube.

N = Combined constant for weight-flow measurement is 68,045 when W is in pounds per day.

D = Inside diameter of the meter tube given in inches.

F_a = Correction for thermal expansion of the primary device (orifice) and is taken as 1.000 so long as the temperature of the mixture at measurement remains between 23 F and 99 F using a steel orifice plate.

F_m = Manometer factor that for a bellows-type meter is 1.000.

F_c = Viscosity factor. Usually this factor is assumed to be 1.000.

F_p = Correction for compressibility of the liquid. This factor is sometimes combined in the definition of G_f.

G_f = Specific gravity of liquid stream at a flowing temperature and pressure as determined by gravitometer readings.

h_w = Differential pressure in inches of water as recorded by a bellows-type orifice meter using flange taps with flowing liquid in lead lines and bellows chambers.

This equation, for simplicity, may be converted to:

$$W = 68,045 \, S \, D^2 \sqrt{G_f h_w}$$

In each of these equations, flow coefficients corrected pursuant to contract or specified conditions and units must be developed. Each coefficient is then multiplied by the square root of the chart differential pressure in inches of water to get the flow rate per hour. Such a rate times the number of hours would give the quantity measured during a given period of time.

Using the orifice meter, as above, or any of the following briefly described liquid measurement methods, it is essential that the fluid must be in a single phase, that is, all liquid at the point of measurement.

Positive Displacement

Liquid measurement by use of positive displacement methods can be very accurate when appropriate corrections are applied. Examples of procedures to be followed are included in API Standard 1101 titled "Measurement of Petroleum Liquid Hydrocarbons By Positive Displacement Meter." Other publications containing information relating to liquid measurements include API Standard 2502, "Lease Automatic Custody Transfer."

In the API and other manuals, it is significantly noted that the equipment requires periodic maintenance and regular proving. The frequency of these operations is dependent on the quantity of fluids being metered. Sampling is critical. Continuous samplers are recommended and should include agitation as a means to enable accurage impurity determination.

Sizing of meters is also critical as are other considerations listed in the API Standards.

Turbine Meter

The turbine meter, described briefly in the coverage of gas measurement above, is very good for liquid measurement if there is no emulsification of the fluid at the point of measurement. The manufacturers furnish calibration and operation data with each unit as well as calculation procedures for gallon or pound units.

Two-Phase Flow

In the above coverage of gas and liquid measurements, emphasis was given to the fact that accuracy suffers if the fluid is not in single phase. However, it is frequently necessary to make measurements for operation and allocation purposes when the fluid is two-phase, that is, both gas and liquid.

If measurement is required of a two-phase stream, certain precautions may be taken to arrive at reasonably acceptable measurements with orifice meters.

1. Keep pressure and temperature as high as feasible at the meter.
2. Use a free-water knockout ahead of the meter.
3. A vertical meter run may improve sometimes the differential-pressure relationship to the volume.
4. Use test data from periodic full-scale separator tests to determine coefficient or meter factor.
5. Connect manifold lead lines to bottom of bellows-type meter with self-draining pots installed above orifice fitting.

NATURAL GAS TESTING

Gas testing is an important phase of the field handling of natural gas. Although not normally considered as such, it actually is a form of measurement—the determination of the liquid hydrocarbon content of the gas. The results of such tests may determine how much a seller receives for his gas or how much of the liquids are allocated to several leases that may be connected to a common system. Royalty payments may also be directly affected by the results of such tests.

Although complete testing is not always done in the field, sampling is. The results of the analysis can be representative of the gas stream being tested only to the extent the sample is so representative. Thus, sampling procedures must assure that containers and testing equipment are purged of all extraneous gases or vapors. The gas-sampling point must be located such that liquids, if condensation takes place, cannot be drawn directly into the sample container. Special equipment must be provided in the event a two-phase stream is being sampled. Leaks in any part of the sampling or testing equipment cannot be tolerated. Care must also be taken to prevent condensation in the lines connecting a sample container to the source of the gas being sampled.

A number of methods are used to test gas depending upon its composition, the gas-sales contract provisions, and the use to which the results are to be put.

Charcoal Tests

This method is largely limited to gas which is low in C_4+ content—usually 1.0 gallon per thousand cubic feet (GPM) or less. The gas is drawn from the stream being tested and passed over activated charcoal. Hydrocarbons are selectively adsorbed to the charcoal with the heavier components replacing the lighter. After the adsorption process is complete, the charcoal is taken to a laboratory where the adsorbed liquid is driven off by heat, condensed, and measured. The results are converted to gallons per 1,000 cu ft of gas. The results obtained from this testing method can be affected substantially by such things as the rate and length of time the gas sample is passed over the charcoal, the quality of the charcoal, and the techniques used in recovering the hydrocarbons from the charcoal. It is used where other methods of testing either give unsatisfactory results or are too expensive.

Compression Testing

This method is used extensively on casing-head gas that has a content of above 0.5-1.0 GPM. The gas sample is compressed to 250-psig pressure in a portable compressor and then cooled in an ice-water bath or in a refrigerated condenser to near 32 F. The liquid so condensed is converted to GPM in the same manner as in the charcoal-test procedure. Compression testing is quite reliable. It is essential to have all equipment in top operating condition if repeatable test data are to be obtained. This is the one form of gas testing that is done completely at the field location.

Fractional Analysis

This method is used in those cases where a knowledge of gas composition is needed. It is also used in testing two-phase gas-condensate streams. In the procedure, a sample of the gas stream is obtained in a metal container and shipped to a laboratory where a fractional analysis is performed. In the case of two-phase streams, it is necessary to separate the gas and liquid in a separator and obtain two samples— one gas and one liquid. The rate of production of each phase is obtained at the time of sampling so that a composite sample can be calculated after each sample has been analyzed.

It is obvious that extreme care must be used in obtaining samples to prevent any contamination. Also, in the case of two-phase streams, precise data on production rates, pressures, temperatures, and so forth are essential.

Laboratory data on fractional analysis will generally show not only the composition in percent of each hydrocarbon present through hexanes or heptanes but will show also the GPM by component (usually beginning with propane) and the heating value of the gas.

Field procedures for the first two testing methods require the use of a test car. Usually representatives of the buyer and seller meet at the test location and check each other, verifying the finally accepted result. The frequency and time of testing is set out in the contract covering the sale. When analyzing is done in a laboratory as in the third testing method, an independent commercial laboratory is usually employed. These firms perform the entire test—sampling as well as analysis.

For more detailed information, refer to the following publications issued by the Natural Gas Processors Association.

AGA-NGPA Code 101-43: Standard Compression and Charcoal Tests for Determining the Natural Gasoline Content of Natural Gas

NGPA 2166-68: Methods for Obtaining Natural Gas Samples for Analysis by Gas Chromatography

NGPA 2165-68: Method for Analysis of Natural Gas Liquid Mixture by Gas Chromatography

NGPA 2161-72: Method of Analysis for Natural Gas and Similar Gaseous Mixtures by Gas Chromatography

NGPA 2265-68: Method of Determination of Hydrogen Sulfide and Mercaptan Sulfur in Natural Gas

APPENDIX A
GAS-FACILITY MAINTENANCE

Generally speaking, there are two main types of maintenance programs: (1) unplanned and (2) preventive. Under the first type, repairs are not made until there is a breakdown or some sort of unexpected shutdown. In the case of a preventive-maintenance program, equipment checks are made on a planned basis with the objective of making needed repairs before a failure or shutdown occurs. Preventive-maintenance programs can be quite simple or very elaborate in accordance with the operators desires. Such planned programs are based on actual performance history, and the keeping of repair and operating records is essential. These records may be relatively simple or very detailed depending upon the elaborateness of the program in effect. As an example, some maintenance programs are designed for major equipment items (i.e., large compressors or pumps), and preventive-maintenance work is done at arbitrarily determined intervals such as once a year or once a month or once a number of operating hours. In such cases, only rather general records of performance are required. At the other extreme, some programs utilize computers that compile and process the operating and repair data needed to provide the basis for the program. In such cases, the records prepared by operating and maintenance personnel must be very accurate and as detailed as the plan requires. There are any number of plans falling in-between these two extremes.

The decision as to the type of maintenance program to be used in any particular operation has to be made by the operator or company involved after some sort of economic evaluation has been completed. In this evaluation, the operator considers production losses due to downtime, operating expense, and maintenance costs (including wages, repair material, cost of special tools, etc.) for several types of maintenance programs to determine which one has the least total cost. Preventive maintenance programs of some sort are selected by most operators because they have determined that repair costs under this

type program are no greater than that for the unplanned type (and often are less in the long run) while at the same time downtime with its attendant production losses is greatly reduced. There are many cases, however, in which it is more economical to let equipment run till it fails and then replace it with repaired or new material that is kept available. Small pumps, magnetos, and spark plugs are examples of such equipment, which is either readily available at stores or can be economically stored on the job site. In reaching his decision in such cases, the operator must take into account the investment cost of having spare equipment on hand, production losses, and the cost of the maintenance program itself. Very sophistocated electronic devices are available to detect potential failures well before they happen. Their use reduces the need to shut equipment down for visual inspection. The operator must consider the cost of the equipment and its operating and maintenance cost in determining whether to use these devices or employ a less elaborate program.

Preventive maintenance is an absolute must for safety and protective shutdown devices and alarms. In order to be assured of reliable operation, this type of equipment must be checked on a very frequent schedule. Unless the operator has a reasonable assurance that it will work when an emergency condition occurs, the money spent for its installation will have been wasted. The lives of operating personnel and the general public may be jeopardized if these devices are not in good operating condition at all times, and the equipment ostensibly being protected is also endangered.

Some of the factors to be considered in making the economic evaluation necessary to the proper selection of a maintenance program are as follows.

Load. Obviously, heavily loaded equipment requires more attention than that which is lightly loaded. A preventive-maintenance program will definitely keep downtime at a minimum and improve production. Electronic testing devices are particularly helpful in such cases since shutdowns for maintenance are made only when potential trouble is indicated by the testing equipment.

Availability of Repair or Replacement Parts and Equipment. At times it takes several weeks to obtain important repair parts. If the investment cost is reasonable, repair parts can be stocked at the site of the operation. Otherwise, frequent checking regarding wear, fatigue, and so forth needs to be made to establish ahead of time when a replacement or repair should be made. Such checking provides the basis for making the repair on a planned basis and at the most opportune time. This should reduce repair part investments, minimize production losses, and perhaps eliminate excessive repair costs that often occur when equipment completely fails.

Standby Equipment. As a facility gets older, it often happens that it becomes somewhat underloaded. The net result may be that the equipment idled by the reduced load, in effect, becomes standby equipment. Any maintenance program should take this condition into account. A much less rigid maintenance schedule can be utilized since production losses should not be such a dominant factor. The installation of standby equipment in new facilities can only be justified if the installation cost can be amortized quickly by savings in maintenance cost plus reduction in production losses.

Life of Operation. Some equipment has little or no salvage value once it has been installed. Buildings, buried pipelines, and so forth are examples of such equipment. If the life of such equipment is approximately the same as that of the project itself, there is little or no justification for maintaining it other than for appearance. Adequate testing procedures (e.g., testing for rate of corrosion in pipelines) are essential, and the life of the project must be established beyond question. Any established program should be reviewed periodically to determine if it should be reduced in scope or discontinued altogether as dictated by the remaining life of the project.

Operating Conditions. Weather, isolation, inaccessibility, and so forth all must be considered in selecting a maintenance program. Moving machinery in dusty areas must have special attention. Rust and corrosion must be controlled in humid areas. Unattended equipment needs to have protective shutdown devices in good operating condition. The spare part supply becomes important where accessibility is a factor.

Environment. Prevention of all types of pollution is becoming more and more important. High operating efficiency as a rule reduces pollutants from equipment such as internal combustion engines, and this can be obtained through good maintenance. Devices installed for the express purpose of controlling pollution need constant attention for proper functioning. Even an occasional spill is no longer condoned by regulatory bodies. The control of noise is becoming necessary, and the maintenance of the dampening devices is a continuous matter. A backfiring engine quickly brings complaints from nearby residents. Maintenance of this type of equipment is not justified economically; it is becoming one of the costs of doing business.

Hazards. Natural gas is safe enough when properly handled. Improperly maintained gas-handling equipment can be a threat to the public, employees, and to the owner's property. This is one of the most important factors to be considered in choosing a maintenance program. Again, economics do not apply directly for there is no way to place a monetary value on a persons's life or well-being. The equipment must be safely constructed, operated, and maintained.

APPENDIX B
NOTES ON GAS-PROCESSING PLANTS

The facilities normally operated in the handling of gas in the field are those required to condition the gas only to make it marketable—removal of impurities, water, excess hydrocarbon liquids, control of delivery pressure through the use of pressure reducing regulators or compressors, and so forth.

Gas-processing plants are usually designed to remove certain valuable products over and above those needed to make the gas marketable, that is, natural gasoline, butane, propane, ethane, and even methane in some instances. In order to do so, plant processes include many of the functions ordinarily performed by gas-conditioning equipment such as dehydration and H_2S removal. Thus, a plant may be considered as another gas-conditioning facility. The equipment used is essentially the same as lease equipment for liquid removal, dehydration, acid-gas removal, and so forth. Plants usually provide fractionating equipment to separate the liquid hydrocarbons recovered into pure products or predetermined mixtures. Where H_2S is removed from the gas, a plant may include facilities to recover elemental sulfur from this impurity.

The decision as to whether or not a plant should be installed is often greatly influenced by the amount of gas conditioning needed. Since condensible hydrocarbon liquids cannot be tolerated by a gas-transmission system, it is mandatory that these products be removed from the gas before it enters the pipeline. A plant can more efficiently remove these liquids and can recover a greater quantity of them than is possible with a smaller lease processing facility. The increased liquid production may well economically justify the entire gas-processing plant installation resulting in a substantially greater money return than would result from simple gas conditioning with the usual field facilities.

Plants are basic to the cycling of a retrograde reservoir since the purpose of the operation is to recover the maximum quantity of liquids while maintaining reservoir pressure. While the use of simple separators and compressors would recover most of the heavier condensate, recoveries can be greatly increased with a processing plant and separation of the products into various components is possible.

Plants may be justified simply for the recovery of certain hydrocarbon components even if those components are only a very small part of the gas stream. As an example, a plant may be installed to process the gas of a transmission system for the recovery of a very small percentage of propane or ethane. The gas volume throughput of a transmission line is usually large, so that the actual gallons of liquid propane or ethane recovered can be substantial. In such cases, large volumes are essential for the project to be economical. The product recovered often has a special value as chemical feedstock. Such plants are not installed to provide conditioning for the natural gas itself.

Plant size is governed by many factors. Quite small plants—20 to 50 MMcf/D—can be profitable when processing the fairly rich gas from an oil field or of the gas from a rich gas—condensate field. Plants of this size are often semiportable and may operate on a largely unattended basis. As pointed out above, where recoveries are small, large volumes of gas normally must be processed. Cycling plants may handle several hundred million cubic feet per day. The processing of gas from several fields for sale is often performed in a single centrally located plant, which again may be very large—perhaps having a throughput of several hundred million cubic feet per day.

Equipment found in gas-plant operations is similar to that found in lease operation with the primary difference being the size and perhaps, the mechanical design of the units. Typical types of equipment found in plants are absorbers, strippers, fractionators, heat exchangers, and aerial coolers or cooling towers.

PLANT PRODUCTS

Plant products are normally considered to be those materials recovered from the natural gas stream that have value in addition to the residue gas itself. The type of operation determines the products to be saved. Cycling plants may produce everything from ethane to naptha or heavy condensate, including motor fuels. Plants processing casing-head gas can produce ethane, propane, butanes, pentanes, and natural gasoline. Plants operating on very dry gas may produce only propane, butane, and natural gasoline. Sulfur, of course, may be produced by plants handling gas containing H_2S. The amount and kinds of products produced in a given plant will depend on composition of the feed gas and the contractual specifications of the residue gas delivered from the plant. Thus, ethane and propane production may be limited in order to maintain the Btu content of the residue gas.

In some new plants, the residue gas is liquefied for transportation where pipelines are not available or possible. This is known as liquefied natural gas (LNG). It is the product of a refrigeration process that requires operations at very low—cryogenic—temperature ranges (about -260 F).

Product specifications and methods of sampling and testing are published by the Natural Gas Processors Association and the American Society for Testing of Materials. Many of the specifications will be found in the Natural Gas Suppliers Association *Engineering Data Book.* Sales contracts for both liquids and gas usually set out the specifications that must be met.

HYDROCARBON RECOVERY PROCESSES

The recovery of liquid hydrocarbons from natural gas in a plant is accomplished by changing conditions of the gas so that the equilibrium between the various components is upset causing some components to condense and others to vaporize in attempting to reach a new equilibrium. The conditions that are changed may be pressure or temperature, it may be the introduction of a different material into the gas stream, or more likely it will be a combination of all three.

The early method of recovering liquid hydrocarbons from natural gas was by means of compression and cooling. Engineers found that by com-

pressing natural gas to higher pressures and cooling it to near ambient temperature, certain hydrocarbon liquids were formed and could be separated from the gas stream. It was noted that this recovery could be predicted using the equilibrium vaporization constants and the natural gas analysis. The compression and cooling process was and is by far the simplest method. However, it is not as efficient as some of the methods developed later. The compression and cooling method was normally limited to cooling by ambient air or by the use of cooling water. A logical development from the conventional compression and cooling method was the use of refrigeration to further reduce the temperature of the gas stream and recover more of the product. The use of mechanical refrigeration systems with ammonia or propane as the refrigerant were the first types of refrigeration used. Of course, the early attempts met several problems associated with hydrate formation. Freezing occurred in the gas chiller and in the separator downstream of the chiller. Injection of methanol or a glycol solution into the gas stream greatly helped this problem. In some cases, there were problems with wax and paraffin formation in the chiller that had to be resolved. In normal plant work, the low-temperature separators usually operate in the temperature range of -20 F to $+20$ F. Refrigeration systems include the use of freon, ammonia absorption, or more recently a turboexpander where a pressure drop is available in the natural gas stream or can be provided economically. Each system is for the same purpose—to cool the gas stream to a temperature that will cause liquefiable hydrocarbons to condense.

The absorption system of hydrocarbon recovery employs the principle of introducing a different material into contact with the gas stream. Absorption oil, having a particular boiling range and molecular weight, is flowed countercurrent to the wet gas in a tray-filled absorber tower. The heavier components of the feed gas condense and flow to the bottom of the tower with the absorption oil. Dry residue gas flows out of the top of the tower. The rich absorption oil is heated in a still (i.e., a fractionating column) where the liquid fractions recovered from the gas are distilled from the absorption oil. The lean oil is returned to the absorber. The recovered liquids are then fractionated into the desired products—propane, butane, natural gasoline, and so forth. By applying refrigeration to both absorption oil and the feed gas, recoveries can be improved. Recoveries of 40 to 60 percent of the ethane and 100 percent of the propane and heavier products are obtained with operating

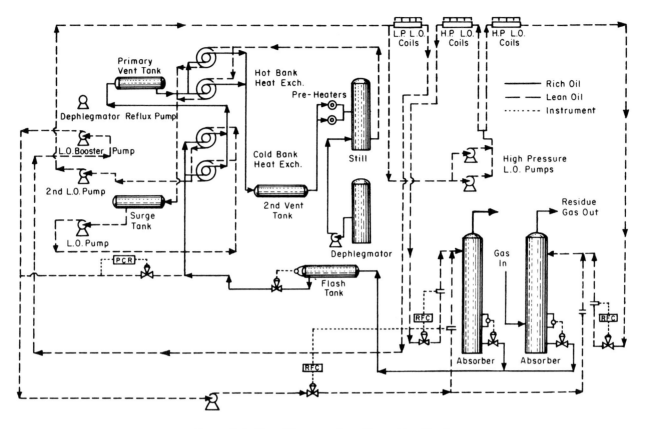

Figure B.1. Flow Diagram of an Absorption Plant

temperatures of −20 F. A flow diagram of an absorption plant is shown in figure B.1.

The most recent development in low-temperature, high-recovery plants is the cryogenic plant that uses the turboexpander. In this type of plant, the gas is expanded through a turbine compressor from which it exhausts at extremely low temperatures in the range −160 to −180 F. At these low temperatures, most of the gas except methane is condensed. The liquids are then fractionated to recover desired products in an ordinary fractionating system. Dehydration is important in any system employing low temperatures. In cryogenic plants nearly 100 percent dehydration is absolutely essential at all times.

ABSORBER AND STRIPPER UNITS

Absorbers and strippers are built much alike. The object of the stripper is to remove something from the liquid stream with gas; in the absorber, liquids are removed from the gas. These may be packed type or tray-type towers. The tray type may have bubble-cap trays, float-valve trays, or sieve-type trays depending

upon the process under consideration. Figure B.2 is a typical packed tower, and figure B.3 is a typical bubble-cap tray tower.

Fractionators and stabilizers are very similar to absorbers and strippers except that the feed point is either near the top or midpoint, and a liquid or liquid-gas feed is made to the tower. Overhead condensers and reboilers are used to control the top and bottom operating temperatures of fractionators. Figure B.4 is a photograph showing several fractionators and stabilizers at a gasoline plant.

Heat exchangers in gas-process plants are usually shell and tube water-cooled units or the air-cooled-type exchangers. Air-cooled exchangers are being used more and more today because of their lower cost, the elimination of water supply and treatment problems, and the ease of maintenance. Shell and tube exchangers require more periodic cleaning to eliminate fouling. In some cases, chemical cleaning is required and even periodic bundle replacement. Air-cooled exchangers are usually provided with temperature-controlled louvers so that the outlet process temperature can be carefully controlled. Recent use of recycle air ducting on air coolers has

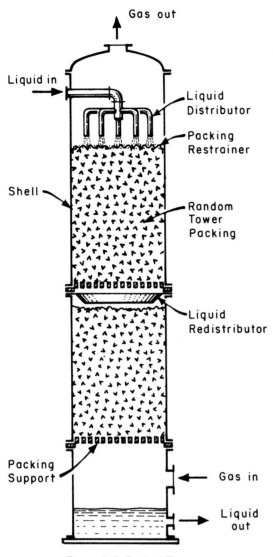

Figure B.2. Packed Tower

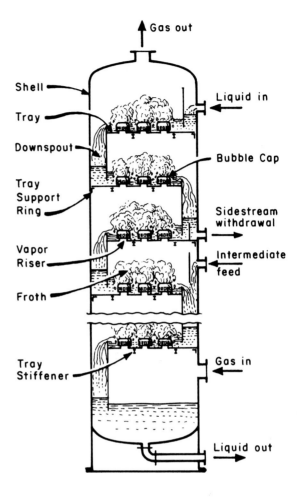

Figure B.3. Bubble-Cap Tray Tower

provided increased flexibility in adapting air-cooled exchangers to process plants for northern climates. Additional temperature control can be obtained with air coolers by using variable speed fan motors, adjustable pitch fans, and variable speed fan drives.

Heat in the process is supplied through the use of indirect heaters, hot oil heaters, salt-bath heaters, and direct-fired heaters.

Figure B.4. Fractionator and Stabilizer Towers